HOW TO DO YOUR DISSERTATION IN GEOGRAPHY AND RELATED DISCIPLINES

Second edition

Tony Parsons and Peter G. Knight

Routledge
Taylor & Francis Group

LONDON AND NEW YORK

First edition published 1995
by Chapman and Hall

Reprinted 1999
by Stanley Thornes (Publishers) Ltd

Reprinted 2001 by Nelson Thornes Ltd

Second edition published 2005
by Routledge
2 Park Square, Milton Park, Abingdon, Oxon OX14 4RN

Simultaneously published in the USA and Canada
by Routledge
270 Madison Ave, New York, NY 10016

Routledge is an imprint of the Taylor & Francis Group

© 2005 Tony Parsons and Peter G. Knight

Typeset in Bembo and Helvetica by
Florence Production Ltd, Stoodleigh, Devon

Printed and Bound in Great Britain by
TJ International Ltd, Padstow, Cornwall

British Library Cataloguing in Publication Data
A catalogue record for this book is available from the British Library

Library of Congress Cataloging in Publication Data
A catalog record for this book has been requested

ISBN 0–415–34154–X (hbk)
ISBN 0–415–34155–8 (pbk)

CONTENTS

FIGURES

TABLES

BOXES

ACKNOWLEDGEMENTS

We thank the many students of Geography at Keele and Leicester whose dissertations provided the basis for many of the examples used to illustrate this book. For the first edition, colleagues from several institutions, particularly Roy Alexander, Tony Budd, Bob Dugdale, Barbara Kennedy, David Pepper, Neil Roberts, John Wainwright and Steve Williams, generously spared time to give us some insight into the role of dissertations in their undergraduate courses. Denys Brunsden and Derek Mottershead were kind enough to read a draft of the manuscript of the first edition and to provide many helpful suggestions that aided us in preparing the final version. We were helped in preparing the second edition by comments from users of the first edition, obtained by our publishers, and by Claire Mercer, whom we thank for helpful discussions and, along with Mike Bradshaw and Angus Cameron, for comments on draft chapters of the second edition. Finally, the front cover of the book shows 'Karoo', a screen print by Dame Barbara Rae, who kindly gave us permission to reproduce it.

PREFACE TO THE
SECOND EDITION

In writing the first edition of this book we hoped that it might prove useful to students who had to write a dissertation, and it turned out that many students found it helpful. This is a good thing, of course, but unfortunately it has meant that the first edition is still being used very widely, even though we have been aware for some time that some parts have fallen out of date, while others could be improved to make it even more useful to even more students. For those reasons, when we came to write the companion to this book, *How to do your Essays, Exams and Coursework in Geography and Related Disciplines*, we and our publishers thought it would be a good idea if we were to revise this book as well. We have taken advice from users of the book (both students and lecturers), and kept the overall format of the book unchanged. It didn't seem to be broken, so we didn't see any reason to try and fix it (sound advice when you are reviewing your own dissertations). However, we have re-written a number of sections and added some new material.

We knew when we wrote the first edition of this book that it was biased towards physical geography. In this edition we have tried to reduce this bias by giving more consideration to approaches to problem solving outside of the hypothetico-deductive model, and more examples from other areas of the discipline. As a consequence of this, we have added some advice about approaches that are common in human geography and in the social sciences generally, and we have tried to make it clear when we are giving advice that is specific to one particular branch of the discipline. Perhaps as a result of writing the first edition, when we've been marking dissertations we have taken greater note of the things we don't like to see. During the decade since we wrote the first edition, we've seen a lot of things that we haven't liked, so we've taken the opportunity to advise against them here!

Some things within geography have changed a lot since we wrote the first edition. For example, the internet, electronic journals and even the virtually ubiquitous use of word processing for student assignments are recent innovations that the old book didn't account for. Our advice about employing a typist to produce your text, for example, now seems a little fuddy-duddy. In the new edition, this sort of fuddy-duddy material has been gracefully retired in favour of more relevant advice about web resources, embedding electronic images within text and issues of that nature. Styles of teaching and examining have also changed, and we have had to take account of that, too. For example, in the first edition we had a number of boxes designed to alert you to some of the searching and difficult questions your examiners could expect you to be able to answer in your viva exam. Vivas are much less common than they used to be, but we've left the viva questions in because they are the same questions that your examiners might think of while reading your dissertation, so you should think about them while you are writing it.

One thing in particular remains unchanged from the first edition: our sincere hope that if you have to write a dissertation, this book will help. Good luck!

1

INTRODUCTION

This chapter explains what the book is trying to achieve and how
you can use the book to get the most out of it.

The aim of this book

The aim of this book is to help you to write the best dissertation you possibly can,
and to help you to get the highest possible mark for it. This book can't write your
dissertation for you, but it can help you to write it to the very best of your ability.
We think you can probably do better than you expect.

In theory, your dissertation should get you one of your highest marks. It's like an
exam where you set the question yourself and have a whole year to find the answer
using as many books as you like and getting help and advice from as many sources
as you can. In fact, many students (and their tutors) are disappointed with the marks
they get for the dissertation. Most students do not do as well as they could; this book
is for them. If you have to write a dissertation, then this book is here to help!

How to use this book

This book will be of help to you from now until the time you hand in your dissertation. Don't just put it away on the shelf after you've glanced through it. Keep it
in the file or box (or pile on the bedroom floor) with the rest of your dissertation
material. As your dissertation progresses you should be able to use the book as a
constant source of advice. The book is arranged a little bit like a workshop manual
for do-it-yourself car mechanics. After this introductory section, the book takes in
turn each stage that you will need to go through in your work – each of the
elements of the dissertation – and suggests ways of tackling the job. It is very much
a 'how to . . .' reference book.

It would be a good idea to read through the whole book quickly right at the
outset. Don't plough through it too diligently at this stage, just amble through it
to see what it's all about. Perhaps read the summaries at the beginning and end of
each chapter. It will help to give you a clear idea of all the trials that lie ahead in
your dissertation. You'll need to know what's involved in the later stages of dissertation preparation even while you are planning the early stages, and if you have a
good overview of the ground, you will find it easier to put each stage of your
work in context.

We've written this book as a linear narrative: going through each of the steps involved in a dissertation in turn. That's because it makes the book easier to use. If you want information on writing up your dissertation, we've put it all together in Chapter 11. But doing your dissertation isn't such a simple linear process, so the first time you need to think about the format of your references, for example, isn't after you've written the whole dissertation. You need to make sure that you have all the information that you will need to format your references correctly as you do the background reading to your study (Chapter 4). Similarly, when you come to write up your methodology (Chapter 11), you will almost certainly be able to use material that you prepared for your research proposal (Chapter 5).

As you start serious work on your dissertation, you will find a chapter of the book dedicated to each of the tasks you have to do. Re-read the relevant chapter as each stage of your work approaches. Many of the chapters include step-by-step procedures that you can adopt to carry you through sections of the project. Throughout the book, generally at the end of each chapter, we have included 'reminders' or 'instructions' of the type: 'when you've read this chapter, do this . . . '. If you are using the book, as we hope, as a manual or guidebook, then following these instructions will ensure that everything is going according to plan and that your work is on target. The idea behind the instructions is that people often find that having specific tasks forces them to focus their attention more directly than having only general advice.

This book does not replace, supersede or in any way supplant your institutional guidelines. Your department will have its own rules and regulations about dissertations, *which you must follow*. If the advice in this book conflicts with the rules of your institution, then follow their rules. Some of the rules relate to issues like the size of your typeface, or the colour of your binding, which clearly won't affect the basis of your work and can be sorted out near to the completion of the project. Other rules might be more fundamental. For example, your institution may have particularly strict rules about how long your dissertation must be, or what type of topic you are allowed to choose. Most institutions allow you to ask your tutor to read a preliminary draft of your dissertation (or part of it), whereas others, strangely, do not. You should check straightaway what your institution expects and allows. Get hold of the official guidelines as soon as possible. Most departments have a formal handbook, which may be a paper document or a web page. If it's a web page, it is probably worth printing out a copy of the key sections to keep with your other dissertation materials. If there are any last-minute changes to the rules, or updates to the advice that you are offered, these will most likely appear on the web page so even if you do get a paper copy, check the on-line version for updates from time to time. It's really important to make sure that you get a copy of the handbook and to make sure you read it very, very carefully. A surprising number of serious problems can arise if you don't follow the rules.

We have tried to make the book as 'user-friendly' as possible, and to leave it up to you exactly how to use it. We've broken the text up into chapters in such a way that you can select only what is relevant to your specific needs at particular times. For example, if you are not doing any fieldwork in your project, then obviously you don't want to wade through sections about fieldwork. Each chapter includes a brief summary at the start, so you can check in advance to see if you need to read the whole thing and you can check afterwards to remind yourself of the key points. We've separated special sections like the 'reminders' out of the main text so you can easily ignore them if you wish, and we've highlighted key issues by the use of short, clearly signposted, sections of text. Each chapter has a short summary at the end to remind you of the main points of what that chapter is about.

You've got enough trouble having to write a dissertation, so we don't want this book to give you any more!

Bear in mind that we've written this book in a pretty informal style, but that your dissertation is a more formal piece of work. We hope that you'll use the book as a source of sound advice, but not copy our informal tone in your own work!

Our friend Eric

When you are working on something like a dissertation, it's good to have a friend to bounce ideas off, to share worries with and maybe to enlist to help with fieldwork. From time to time in this book we'll mention our friend Eric; he represents the kind of character it's sometimes useful to have around. We'll use Eric to throw up alternative ideas about what we've said, and if we want to refer to whomever it is that you have around to talk to, we'll call that person Eric too! He's a sort of universal side-kick. If he gets on your nerves, well, people can be like that; your best bet is to listen to what everybody says, but to remember that it's your dissertation, and that you have to make the decisions.

Eric says: **You need all the help you can get!**

Dissertations: what this book is all about

Most undergraduate courses in geography and related disciplines include a dissertation as part of the assessment. The precise nature of the dissertation varies a little among institutions (see Chapter 2) but the basic requirements are much the same wherever you are studying. Essentially, the dissertation is a project of your own. You do some research and you present a report of that research. Your research, as presented in the report, is assessed.

Whether you think of yourself as an arts, humanities or science student, and whether your dissertation deals with a topic in human geography or in physical geography, you should think of your dissertation as a 'scientific' report. Science is all about investigation, understanding and communication, and is not confined to test tubes and Bunsen burners. Your dissertation involves the acquisition and communication of knowledge; science, in its broadest sense, is exactly that. Your dissertation should be an exercise in the 'scientific' virtues of organization, precision and clarity.

The first important point for us to make before you start is that there are good ways and bad ways of going about research, and there are good ways and bad ways of writing reports. It is very important that you find a good way. You need to go about your research efficiently, and you need to present your report effectively. Unfortunately, it isn't always easy to sort out good approaches from bad when you are planning your first piece of research, and there is plenty of middle ground to swallow up mediocre students and turn them into 'thirds'. The dissertation element of a degree course is a test of your ability to negotiate the unfamiliar territories of research design and scientific writing. We hope this book will serve as a guide.

Chapter summary and conclusion

- The aim of this book is to help you produce the best dissertation you can.
- Use it as a guide and manual throughout your dissertation.
- Use it in conjunction with the guidelines issued by your own institution.

What to do after reading Chapter 1

When you've read Chapter 1, before you go any further with your dissertation, make sure you have a copy of all the instructions and guidelines about dissertations issued by your institution. Your department probably has some sort of 'dissertation handout' designed for students. If you haven't received one yet, ask for one. Most departments have one member of staff with special responsibility for the administration of dissertations. If in doubt, ask the departmental secretary, the departmental administrator or one of your own tutors. Before you start your dissertation, you must find out exactly what the rules are and what is required of you. When you get the handout, don't just file it away. Read it carefully, refer back to it as your work proceeds and check your dissertation web page for updates to be sure that you stay on target.

GET THAT HANDOUT NOW!

2

WHAT IS A (GOOD) DISSERTATION AND WHY DO I HAVE TO DO ONE?

This chapter explains the essential elements of a dissertation and what distinguishes it from other work you have done; tells you what to look for (and what your examiners will be looking for) in a good one; and shows what you can expect to get out of doing one.

What is a dissertation?

A dissertation can be defined as a report on an original piece of research. This definition contains three pieces of information that are important to you.

First, your dissertation is 'a piece of research'. This means that in the course of doing your dissertation you will need to try to find out something. This requirement is very important because it defines how you will need to go about your dissertation. It won't be sufficient to write what you already know about a topic, or even to write what you and everybody else knows. It isn't, therefore, a discourse or just a longer-than-usual essay. It is fundamentally different from most of the things you will have been required to write during your undergraduate course and is significantly different from any project that you may have done at school. Not surprisingly, it requires a very different approach. This may all sound daunting, but it isn't. Challenging, perhaps, but not daunting. Remember, you only have to try to find out *something* – *anything*. It doesn't have to be important or earth-shattering for you to produce a good dissertation. Naturally, if you do find out something important it will help your dissertation a lot. On the other hand, a good report on a minor piece of original research can gain you a first-class mark. (One of the decisions you will have to make will be how big or important a topic you investigate – see Chapter 4.) Equally, you don't actually have to *succeed* in finding out anything. Many students worry when they discover that their cherished hypothesis comes to nothing and their results support what everybody thought all along. Although this may be disappointing, it needn't affect the quality of your dissertation. We'll have more to say about this in Chapter 10.

Second, the research has to be 'original'. Most obviously, this means that your dissertation has to be your own work and not copied from somebody else. That's plagiarism and your examiners will deal it with very harshly. Plagiarism covers everything from lifting your whole dissertation from the work of somebody else to failing to acknowledge a source of data. We'll say a little more about this, and its consequences, in Chapter 12. The need to be original also means that you can't just repeat

5

a piece of work that somebody else has done. You have either to do something new or take a new approach to something old. Again, this might seem a bit daunting. Here you are, an undergraduate, expected to find out something that the collective endeavours of eminent academics have so far failed to discover. But the world is full of problems to investigate and those eminent academics didn't become eminent by investigating the little ones. You're unlikely to become famous for demonstrating that 50 per cent of the shoppers in your local supermarket live within a five-mile radius and that they visit the supermarket three times a month, on average. On the other hand, you might get a good mark for a dissertation that demonstrated such things. We'll have more to say later about the type of problem you investigate. At this point, it's just important to realize that it isn't that difficult to be 'original'.

Finally, your dissertation is a 'report'. Because you have carried out a piece of original research, you know something that nobody else knows. Your dissertation is no more than telling people what you have found out. The people whom you are telling, however, (your examiners) are difficult to convince. You can't simply tell them what you have found out and expect them to believe you. You have to demonstrate that the methods you employed were sound and that your findings are reliable. Basically, you can think of your dissertation as the recipe for finding out what you have found out. If your examiners were to follow this recipe, they too would reach the same conclusions as you have. They need to be convinced this is so (without actually having to repeat your investigation).

What makes a good dissertation?

A good dissertation is one in which a soundly constructed and executed piece of research is reported on in a clear and logical manner. Your examiners depend on what you have written to decide whether or not what you have done really is soundly constructed and executed. So it is vital that the report that you write is well-structured and clearly written, i.e. easy to follow. Your reader (examiner) wants to know what you've done and how you've done it. Tell your story clearly. This may not be easy. You may be unfamiliar with writing such a long document. More commonly, your own close involvement with your dissertation may make it difficult for you to realize just how little the reader will know of what you are doing, and to appreciate the order in which information needs to be conveyed. But it cannot be overstressed that even the most mediocre piece of research, clearly presented, can come over quite well. Equally, the most dazzling discovery can be destroyed by poor presentation.

A good dissertation is a clear report on a well-designed piece of research.

Obviously, in writing clearly you will expose the weaknesses of your study just as much as the strengths. This shouldn't worry you too much. First, your examiner will find them anyway and, second, in writing clearly, you may see them first! You may not be able to do anything about them at that stage, but you might be able to put up some justification or explanation for them. Better that you should know your weaknesses than, say, turn up to a *viva voce* examination with confidence and have your examiner say 'Of course, I'm sure you realize now that the methods of data analysis you employed in your dissertation were flawed'. Apart from the quality of the writing, there are many other qualities that distinguish good dissertations from others. All these qualities will be used by your examiners in deciding what mark to award your dissertation, so it is a good idea to know what they are and what your examiners will be looking for. Box 2.1 summarizes the qualities of a good dissertation. As Box 2.1 shows, how you go about achieving some of these qualities is dealt with in some detail in later chapters. For these qualities we'll only have a little to say here. For those not covered in later chapters, we'll say rather more.

Box 2.1 The qualities of a good dissertation

- A good problem (see Chapter 4)
- Set in its scientific, or intellectual context
- Logical research programme (see Chapter 5)
- Clearly defined methodology
- Adequate and appropriate data analysis (see Chapter 7)
- Clear separation of results and their interpretation
- Well structured and clearly written (see Chapter 11)
- Intellectual achievement
- Correct conclusions
- Well presented (see Chapter 11)

A good problem

A good problem is one that you will be able to tackle with the time and resources available to you, that lends itself to exploiting any particular skills, interests or expertise that you have and which can be set into a wider context of intellectual enquiry. This is a tall order! Identifying a good problem is perhaps the most difficult task you will face and we will try to give much more detailed advice on how to go about it in Chapter 4.

Scientific or intellectual context

Your examiners will want to know why you carried out your particular piece of research and how it fits into existing knowledge. So a good dissertation will be one that contains an introductory chapter in which the background to the study is presented. Typically, this background will have the form: 'People have studied X since the pioneering work of Jones (1854). In this study Jones argued that . . . More recent investigations by Smithson (1985) have suggested that . . . However, it can be argued that . . . and hence the purpose of this dissertation is . . .'. A brief example of how such a context is established is shown in Box 2.2.

Box 2.2 An example of some background to a study

Much of the research into limestone pavements has been concerned with their formation. In north-west England the major area of controversy centres upon the role
of the Devensian glaciation. Clayton (1981) argued that the pavements of north-
west England were largely due to stripping of weathered limestone by glaciers
down to relatively little-weathered bedding planes. However, Pigott (1965) claimed
that grikes developed beneath a deep weathering mantle during interglacial
times. Williams (1966), on the other hand, believed that the grikes pre-dated the
Devensian glaciation and survived through it. Parry (1960) presented a more complex view and proposed the existence of two types of grike. The first type consists
of crevasse-like clefts that are long and often curved. These grikes, he argued, are
solution-widened joints. The second type of grikes consist of smaller and less
regular furrows that are entirely solutional in origin and are of recent origin.

In a quantitative analysis, Rose and Vincent (1986) measured grike width on
three limestone pavements and claimed that the distribution of grike widths indicated that the grikes were of two ages, thereby supporting Parry's view. This
conclusion was based upon measurement of grike widths to the nearest
millimetre. Notwithstanding the obvious benefits of a quantitative analysis of
grikes to test the hypothesis of two grike populations, the measurements undertaken by Rose and Vincent seem questionable. Grikes seldom have a sharp edge
so that determination of their widths is, necessarily, subjective. Furthermore, grike
width shows great local variation. A preliminary investigation for this project
showed that grike width could vary by up to four times within a metre stretch.
A more appropriate index of grike type might be their depths. Grike depths are
both much larger in general, so that any subjective error in measurement will be
proportionately less significant, and are less subject to great local variation. The
aim of this dissertation is to discover whether grike depth is a better indicator
of grike age than is their width. More specifically, it will examine whether or not
measurements of grike depth support the claim of two types of grikes.

This type of background has two aims. First, you need to demonstrate to your
examiner that yours is the sort of dissertation that he/she is going to want to read.
Examiners like to read good dissertations, so this is an early opportunity to demonstrate that yours is one of these. You do this by demonstrating your expertise in the
topic area. You demonstrate your expertise by showing that you know what has been
done before and is currently being done in this field of research. The chances are that
at least one of your examiners will be knowledgeable in this field and so will also
know this background. It is essential that you convince this person, in particular, that
you are somebody whose work is worth reading. All of the major, and some of the
minor, contributions to this field should be referred to in this background. It is impossible to specify how large this background should be because it will vary enormously
from one field to another, but reference to a couple of dozen papers would be a reasonable average. Second, having convinced your examiners that you are the sort of
person whose work is worth reading, you now need to entice your reader to *want* to

read further. At the end of the background to your study, the reader should be intrigued by the apparent paradox or counter-intuitive observation or gap in understanding that you have identified. Having awakened curiosity in your reader's mind, all you have to do now is to satisfy it.

Research programme

Your research programme is how you go about solving your problem. There are many ways you could do this. Some will be so poor that they could not possibly lead to a solution to your problem. For example, suppose you have decided to investigate the relationship between soil physical properties and slope gradient along a series of hillslope profiles. Because of the site you choose to study, it turns out that all the upper parts of your profiles are on one rock type and all the lower ones are on a second rock type. The existence of this lithological difference is a fundamental flaw in your research design so that no matter how well you do everything else, you cannot possibly solve the problem you set yourself. You may, however, be able to solve a different problem. Making sure that your research programme will lead to solutions to the problems/answers to questions you are investigating and not to the solutions/answers to a different set of problems/questions is important and we will give a lot of attention to it in Chapter 5.

Methodology

This is no more than the way you set about doing your dissertation. Again, think of your dissertation as a recipe. Look at the two examples in Box 2.3 on p. 10 describing how measurements were made of path width and gradient. Clearly, if you wanted to carry out a comparative investigation to the one being discussed here, you would find it much easier to do so if the dissertation followed Example 2 rather than Example 1. Whether or not you think this is a good way of determining path width and gradient, you are in no doubt as to how it was done in this particular study.

Data and data analysis

Your research programme deals with what you need to do to solve your problem and your methodology deals with how you do it. Data and data analysis are concerned with the type of data you will need to collect, the amount and what you do with your data once you have collected them. If you are making measurements, you will need to show that you have measured things to the appropriate level of accuracy and that you have obtained sufficient data for the type of analysis you have undertaken so that the results are valid.

Separation of results from their interpretation

What you find out and what you make of it are quite separate things. So, in the example given in Box 2.3, the author may have found out that all paths in the area

> **Box 2.3** Examples of descriptions of methodology. Example 2 is more helpful to the reader than Example 1
>
> **Example 1**
> To determine the differences in path width and gradient, measurements were made of width and gradient at several locations on each path and the averages of these measurements were used as the width and gradient for that path.
>
> **Example 2**
> To determine the differences in path width and gradient, a representative section of each path was identified. This section, which was 2m long, was chosen at a point where the path was straight, had a uniform gradient and where its direction was unaffected by obstacles. Within this section five readings of width were made at 0.5m intervals. Where overhanging vegetation obscured the edge of the path, the measurement was taken to the point on the path vertically beneath the overhanging vegetation. The average of these five readings was used as the width for that path. Path gradient was measured using an Abney level as a single value for the full 2m section.

studied had the same width even though their gradients differed. He/she might even have used this discovery to argue that, because path width can be used as a measure of path erosion, path erosion is unaffected by path gradient. The inference (that erosion is unaffected by gradient) that is drawn from the discovery (that all paths have the same width even though their gradients differ) is achieved via an argument (that path width can be used as a measure of path erosion). You may disagree with the soundness of the argument, and hence the inference, but that does not affect the validity of the discovery, which depends for its reliability on the methods used to measure path width and gradient. On the other hand, you might disagree with the inference even though you agree with the argument linking path width to path erosion, because you are unhappy about the way path width was measured. You might think that it would have been better to measure to the edge of ground vegetation cover rather than to the point vertically beneath overhanging vegetation. It is very important that the reader of your dissertation can separate out the stages you have gone through in reaching your conclusions and be able to evaluate each of them individually, and it is therefore essential that you present them as separate and clearly identifiable sections of text.

Structure and writing

As we've said before, your dissertation is really like a recipe for solving the problem you have set. Just like a recipe, it's important to get things in the right order so that your examiners can follow through the logic of the study. Likewise, you need to write clearly and unambiguously. We'll give greater detail on this aspect of your dissertation in Chapter 11.

Intellectual achievement

In assessing the intellectual achievement of a dissertation, the examiners will be looking for evidence of capabilities for reasoning, analysis and synthesis. In many dissertations the results of data collection and analysis are simply presented almost without comment. Sections of your dissertation that read something like that shown in Box 2.4 should be present if you are to do well according to this criterion.

Box 2.4   Signs of intellectual achievement in a dissertation

The results of Table 2 suggest that either . . . or . . . is the explanation for the observed difference in the average age of shoppers on different days of the week (Table 1). However, the evidence from Table 3 indicates that . . . cannot be the explanation. This leaves . . . as the only plausible reason for the effect of the day on the age of shoppers.

Putting together the results of Tables 1–6, a broader pattern of shopping behaviour begins to emerge. This behaviour is summarized in the causal linkage model of Figure 9. Although this model can only be tentative on the basis of the available information, it does provide a conceptual framework within which an understanding of shopping behaviour in Stoke-on-Trent may be achieved.

Conclusions

A good dissertation is one that concludes, rather than just stopping. In the conclusion you should summarize your findings and set them in the wider intellectual context in which your dissertation was initially set. So, you might be able to refer back to your introductory chapter and the existing literature on the subject and show how what you have found out fits into/contradicts/supports the earlier literature. A conclusion to the dissertation that began with the background shown in Box 2.2 on p. 8 might look like that shown in Box 2.5 on p.12. The conclusion is also the place to suggest further lines of research or alternative approaches that might be taken to solving your problem. This is particularly useful if your dissertation leads to surprising results, or to no results at all.

Presentation

Your examiners will want to see a well-presented dissertation. They will want to see evidence that you have taken trouble over it and that you have the skills and knowledge to prepare a professional-looking report.

The quality of your presentation can affect the mark you obtain for your dissertation by as much as 20 per cent.

> **Box 2.5 Concluding material to the dissertation that began with the material in Box 2.2**
>
> This study has reported on measurements of grike width and depth on eight limestone pavements, including the three that were examined by Rose and Vincent (1986). The study confirms the preliminary observations of great variability in grike width and, on the basis of comparisons between such measurements made by three independent observers, shows that the measurement of grike width is very subjective. In comparison, it has been shown that grike depth shows much less variation and, using the same three independent observers, is much less susceptible to operator variance. Likewise, the determination of grike age using depth seems to give much more clear-cut results than when these determinations are made from width measurements. It is concluded that grike dimensions can provide a useful tool for identifying the erosional history of limestone pavements (as Rose and Vincent (1986) argued) but that this tool is much sharper if it relies on depth measurements.

If you have taken trouble over the appearance of your dissertation they are more likely to believe that you have taken trouble over things that are less easy to check on. For example, you may say that you obtained readings of stream discharges every 30 minutes. The examiners have to take your word for this. So if you fell asleep in the sunshine or went off to the pub for a leisurely lunch break with the result that some of your readings were taken late, the examiners won't know. If it turns out that in your dissertation you refer to Table 6 when you mean to refer to Table 5, that the pages are not in order, that there are typographical errors, and some of your stream discharge data look a bit peculiar, the examiners are likely to wonder about the accuracy of your data collection. On the other hand, if your dissertation is immaculate and there isn't a single spelling mistake but some of your stream data are peculiar, the examiners are more likely to believe that they are not a result of sloppy work on your part but that they represent real variations in the discharge of your stream. Unlike the marks that may be awarded for intellectual achievement, those for presentation are easy to earn. So don't miss out on them. If you put your dissertation through a spelling checker, make sure it spells in the version of English (UK/US) that your institution uses. If possible, use a grammatical one so that more errors are picked up. But a spelling checker can't do everything. It is unlikely to know that you mean 'effect' when you have written 'affect'. Also, check the final document. Don't assume that it will look OK on paper just because it did on the screen. Printers often do some inexplicable things at the last minute.

But a well-presented dissertation can't make up for sloppy work. It can make it look good, but your examiners will have seen a lot of dissertations. They will be reassured by an attractive document, but seldom fooled!

Why do I have to do a dissertation?

Your dissertation is not simply a torture dreamed up by your tutors to keep you busy. It has very specific educational aims and there are important benefits to be gained from doing it well. Knowing beforehand what these aims are and how you might benefit from your dissertation may give you some reason to do it well and to put effort into it.

Aims of a dissertation

The aims behind requiring you to do a dissertation are to train you to be able to conduct an investigation on your own (but with some guidance) and to report on that investigation. There are several steps in conducting and reporting on such an investigation, and subsequent chapters of this book will deal with these steps in detail. Here, we're more concerned with the reasoning behind doing a dissertation at all. The first principal aim is linked to the fact that there needs to be a group of people in the world who find things out and that you (when you become graduates) are quite likely to find yourselves in, and be expected to be able to contribute to, this group. Also, if society, civilization and science are not to proceed up a series of false tracks and blind alleys, those who do claim to have found things out need to be subjected to scrutiny. People who have tried to find things out themselves are likely to know the pitfalls and so can more easily identify shortcomings in the apparent findings of others. One aim of your dissertation is to make you competent in evaluating the results of others who claim to have found things out, as well as becoming competent in finding things out yourself.

As well as doing the research, you have to report it – that is, write your dissertation. The second main aim therefore is to train you to convey information. Again, there is a need for people to convey information effectively. Your dissertation project aims to teach you the skills to enable you to do that. You will be familiar with the frequent complaints about official forms that are unintelligible to the average person. They are examples of poor communication. An aim of your dissertation project is to train you so that you don't end up as an author of such an unintelligible form. There is no reason why any piece of information cannot be expressed in a manner that is intelligible. As the physicist Richard Feynman is said to have once remarked, when asked to prepare a freshman (first-year) lecture on a particular topic, 'I couldn't reduce it to the freshman level. That means we don't really understand it' (Gleick 1992, p. 399).

Almost all people in any profession need to be able to communicate effectively. Whatever you decide to do once you graduate, effective communication is likely to be a valuable skill to have acquired.

In many ways, you can think of your dissertation as the culmination of your undergraduate course. It's where you bring together all the skills you have learnt in the course of your degree programme.

Benefits of undertaking a dissertation

If you are successful in your dissertation (and obtain a good mark for it), it will indicate that you have succeeded in meeting the aims. You will have become competent at all the steps that are discussed in the remainder of this book. What will that do for you? Who is going to be impressed if you shout out 'I wrote the best dissertation of my year'? Well, you are, for one. Don't underestimate the confidence-boosting benefit of having done something successfully, particularly if you started out thinking that it was unfamiliar, frightening or difficult. If you tackle successfully something about which you had doubts regarding your own ability, you will approach later, seemingly difficult, problems with a bit more confidence.

Your dissertation can:

- train you in research design and execution
- train you in project management
- train you to communicate
- give you self-confidence
- improve your degree class
- help you to get a job.

But you (and your grandmother) aren't the only ones who may be impressed. Once you graduate you will want/need to do something else. Maybe you will decide you'd like to undertake postgraduate training, or maybe you will look for a job that pays real money. Either way, a good dissertation can help. Usually it is carried out and submitted well ahead of finishing your undergraduate course. It is therefore available to use to impress those who might employ you or offer you a place on a postgraduate course. So keep a copy of it and take it along to show to anybody who indicates the slightest interest. If it's very good, make sure you show it even to those who express no interest!

Postgraduate courses involve more research-based activity than undergraduate training. So your dissertation is the best guide as to how well you might do on such courses. Likewise, an employer might expect to be able to give you tasks to carry out and report back on what you have found out. Your dissertation will be a useful guide as to how well you might cope with such tasks.

Finally, remember that in many cases you will want to use your tutor as a referee in your applications. Usually, by the time such references come to be written, your dissertation is well under way and is quite likely to have been submitted and even marked. This will give your tutor something tangible to say about you. By doing well in your dissertation you can make sure that the 'something' is to your advantage.

A prize-winning dissertation?

A lot of people (not just us) think that dissertations are important. To encourage students to write good ones they have instituted prizes for good undergraduate dissertations. The prizes are a long way off winning the National Lottery, but they're nice to win. Some sources of prizes are:

- The Royal Geographical Society
- The British Geomorphological Research Group
- The Remote Sensing Society
- The Institute of Environmental Sciences
- Many universities have one or more prizes for geography dissertations
- The Developing Areas Research Group.

Some of these prizes are restricted to particular parts of the discipline, but others will be open to any type of dissertation. If you are eligible for a prize and your tutor thinks you are in with a chance, it's quite likely that your department will enter it on your behalf, but there's no reason why you shouldn't initiate the process.

Chapter summary and conclusion

This chapter has explored the things that are important about the definition of a dissertation; described the qualities that go to make up a good dissertation; and demonstrated the aims and benefits of undertaking one.

What to do after reading Chapter 2

Your institution will almost certainly have copies of dissertations submitted in the recent past. Look at some of these. Try to see what is good about them; and what is bad. Imagine yourself as an examiner and see what you would comment on. Write these things down for later reference.

3

WHEN SHOULD I START AND HOW LONG WILL IT TAKE?

This chapter describes the stages in the life of a dissertation (all the things you have to do), and how to allocate your time sensibly to each task.

Introduction

You should start now: it will take longer than you think. One of the most common reasons for students writing disappointing dissertations is that they underestimate how long things will take, and end up in a last-minute panic having to rush, or even miss out, important parts of the job. There are lots of separate tasks involved in producing your dissertation, all the way from deciding what kind of topic you will do to checking the finished article one last time and handing it in. If you want to do as well as you possibly can, then it is essential that you know exactly what needs to be done, that you know how much time needs to be allocated to each task, that you know when you will be in a position to work on each task, and that you plan your schedule of work accordingly. The aim of this chapter is to help you to organize your time as well as possible, and to avoid unnecessary panic and consequent disappointment.

The importance of a timetable

Most institutions allow about a year to do the dissertation, and you may have been given as much as 18 months between the time you had the dissertation explained to you and the time you have to hand it in. Compared to the deadlines you will be used to for your tutorial essays or practicals, the dissertation stretches unthreateningly into the distant future. You've probably never been given so much time to do a single piece of work, and it may seem unnecessarily generous. You might think it's just an administrative convenience of some kind for the department to stretch out the dissertation, or some archaic quirk of the academic system. You're wrong. The length of time that your institution allocates to the dissertation is carefully thought out and based on the length of time that it takes to do the job properly. Depending on the specific requirements of different institutions this is, indeed, about one year.

There are lots of separate tasks involved in producing a dissertation. Some of them can be done simultaneously; others need to be done sequentially. For example, you might be able to start writing your introductory chapters at the same time as you are administering your postal questionnaire, and it might be essential to collect your river velocity data at the same time that you collect your suspended sediment data, but you certainly don't want to collect any data until after you've established the subject of your project, and there's no point printing up your postal questionnaire until after you've analysed the results of your pilot study. If you do things in the wrong order, or if you try to do two jobs at the same time that should be done one after the other, you'll get yourself in a right pickle. Worse still, try as you might to cover your tracks later, the examiner will be able to spot that you got in a pickle, and won't be impressed. So, it's important to know what things you need to do, and what order to do them in. As we said in the last paragraph it's also important to get the timing right, so you need to know how long each job, or each group of jobs, will take. Clearly, you need all of this information right at the outset, before you commit yourself to your programme, or else you won't even know where to begin. In other words, you need to worry about this now. The best way to sort all of this out is to make a timetable.

Institutional guidelines and 'the deadline'

Your institution will have its own specific guidelines and regulations about the dissertation. In most institutions these are put into some kind of handout, and may also be available on your institution's web site. This should include information about how long you should expect to spend on the dissertation, when you might reasonably expect to begin the work and, most certainly, a specific date by which you must have the work completed and handed in: the deadline. You need to know what your institution's regulations are, and you must know what the deadline is. If you don't already know, then go and find out. You can't organize your time or draw up a timetable if you don't know the deadline. All your planning depends on that date.

The deadline is the date beyond which you must not go. There will be a severe penalty if you don't get the dissertation in on time. Some institutions simply do not accept late work, and you will score zero, 0 per cent, nothing. This is a very bad thing. In some institutions it means you cannot pass your degree. Other institutions employ systems involving deductions of marks for every hour or day that the dissertation is late. Whatever system your institution employs, you will almost certainly find that being late is a disaster, and you must avoid it at all costs. Meet the deadline.

If there is any legitimate reason that you might not be able to meet the deadline, such as illness, then consult your tutor at the earliest possible time to see what arrangements can be made. You may be granted an extension, or given some allowance in the assessment. Consult your departmental guidelines, and discuss your situation with your tutor, and do so well in advance.

Your institution may also have intermediate deadlines along the way. You may, for example, have to submit a literature review a few months after you start. There may be scheduled appointments with your tutor that you are required to keep. Make a note of these intermediate deadlines.

17

Constructing a timetable

At this stage you should know at least three things: when the deadline is; how long there is to go between now and then; and how much work, if any, you have already done on the dissertation. Now you are in a position to start putting together a timetable.

Constructing the timetable is a valuable exercise in itself, as it forces you to recognize all the different jobs that you will have to do. You will also have to think about what each job will involve, how long it will take, and how it will fit together with the other jobs.

Don't be afraid to go through a lot of scrap paper as you try to get all this stuff out of your head and onto the page. A couple of hours thinking and scribbling at this stage will save you a lot of grief later on. The exercise will also show you something that will crop up over and over again as your project proceeds, namely that your work will not necessarily develop in a linear fashion; as a new idea occurs (or some new disaster befalls your plans!) you will find yourself going backwards and forwards through your work readjusting things to keep everything straight. Don't worry, that is supposed to be part of the game; one of the things that the whole business of doing a dissertation is supposed to teach you to deal with.

A do-it-yourself timetable kit

It is very easy to start building your timetable. You can do it on paper or you can do it on a spreadsheet on your computer. These instructions assume you're sketching it out on paper but it really doesn't matter.

1   Take a clean sheet of paper and write at the bottom: 'Hand in dissertation' and the date of your institution's submission deadline. At the top of the page write: 'Start work on dissertation' and the date when your institution explained the requirements of the dissertation to you.

All the different tasks that you need to do (the reading, the planning, the fieldwork, the labwork, the writing and all the rest) then need to be fitted into the blank space in the middle of your page. Straight away you can:

2   Add the months down the side of the page between 'Start work' and 'Hand in'.
3   Add an arrow with the word 'Now' at today's date.

Eric says: Build your timetable into your on-line calendar or personal organizer!

If looking at this book and thinking about this timetable are the first things you've done towards the dissertation (which would not be unreasonable!), then your 'Now' will be immediately underneath your 'Start'. If you've already spent some time working at it, then you might be able to put a few things above the 'Now' to show what you've done so far. For very many students the timetable at this stage looks like Figure 3.1, with 'Start' at the point where the department gave the introductory lecture, or gave out the dissertation guidelines, then a long gap of 'did nothing' before the 'Now' some way down the page. If your timetable looks like this, you will recognize already that you have less space than you might like into which to fit your tasks.

4 Screw up this first version of the timetable and move on to version two (Figure 3.2), with 'Now' at the very top of the page, 'Submission deadline' at the very bottom, and an encouragingly expansive clean space in between! If you're building your timetable on a spreadsheet you can just delete a few of the redundant rows from the top to achieve the same result.

Now all you have to do is fill in the gaps. Read on!

Fitting the jobs into the timetable

As this point you need to think carefully about what exactly lies ahead and how long it will take you to do it. We say 'you' deliberately. Everybody works at their own pace, and you need to produce a schedule that you can achieve; not one for your friends, your tutor or the mythical 'typical student'. Different projects will

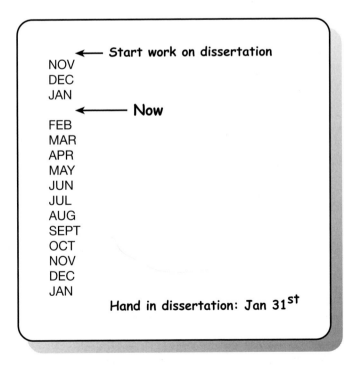

Figure 3.1 First draft of a dissertation timetable.

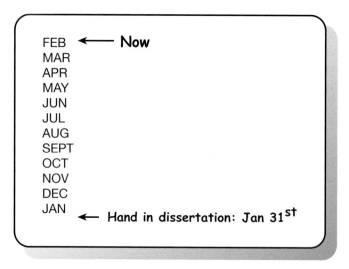

FEB ←— **Now**
MAR
APR
MAY
JUN
JUL
AUG
SEPT
OCT
NOV
DEC
JAN ←— **Hand in dissertation: Jan 31st**

Figure 3.2 Second draft of a dissertation timetable.

involve different tasks, but the basic elements are universal. Looking through the later chapters of this book will give you more detail about what is involved, but for the purpose of timetabling you can start with a fairly simple list such as that in Table 3.1. For the purposes of your particular project you may need to omit parts of this list, or add new headings to it, but this is a good list to start working with.

Table 3.1 An example of the sort of dissertation schedule that can be produced by slotting the jobs to be done into the time available.

Job no.	Job Description	Relative time allocation	Schedule
1	Establish schedule and produce timetable	–	Jan
2	Decide what to do it on: lots of reading and thinking. Specify the 'Topic', 'Problem', 'Question'	2	Jan/Feb
3	Literature search: read, think and revise (2)	1½	Mar/Apr
4	Research design: how to find out what you asked in (2), bearing in mind what you learned in (3)	1½	Apr/May
5	Data collection: fieldwork, questionnaires, lab experiments, getting what (4) required	1	June
6	Data analysis: lab work on field samples, statistical treatment of survey results, etc.	1	July
7	Interpretation: figuring out what (5) and (6) tell you about the question in (2)	1	Aug
8	Writing: write the report, draw the figures, check the drafts, etc.	2	Sept/Oct
9	Preparing the report: print final version, attach inserts, binding	1	Nov
10	Submit report	–	Dec

Some of the jobs that go onto the timetable will have to be done at specific times, and these are the easiest, and the most important, to fit onto your timetable straight away. For example, your institution might insist that certain forms are filled in at certain times, or that research proposals are presented to tutors before a certain date. Your institution's regulations might include specific limits on the latest time at which you will be eligible to apply for financial support for the dissertation, or the last date at which tutors will be prepared to read drafts of your chapters. Equally, your domestic circumstances can place constraints on the timetable; your fieldwork period might be constrained by availability of familial accommodation, or of field assistance, or by the dates of a family holiday. All of these constraints need to be built into your timetable at this stage, along with term and vacation dates, examinations and other commitments. To avoid making your page too messy, things that aren't directly related to the project can be put in a column down the right-hand side of the page.

5   Add the dates of institutional timetable requirements such as term/vacation dates and things like: 'Hand in proposal form to tutor before this time'. Also add other inflexible dates such as: 'Holiday in China with Eric'; or 'Two weeks of exams for car-maintenance night class'; or 'Six weeks for summer-vacation job'.

It will become clear at this point that there are chunks of time when work on the dissertation will be impossible. For example, you will not make much progress on your study of cinema catchment areas in Milton Keynes while you are in China with Eric.

6   Shade in the areas of the timetable when dissertation work is not possible.

You can also be fairly confident that 'unavoidable constraints' (disasters and emergencies) that you didn't anticipate (never dreamed of) will arise. You will lose all your field notes; your computer will crash and eat your disk; the dog will eat the back-up and the back-up back-up will go into the washing machine in your shirt pocket; you will be sent to jail for a fortnight; loads of good stuff will come on the telly; Daddy will insist that you all go on a skiing holiday for three weeks; the central pillar of your statistical analysis will crumble and the whole idea of your project will be rendered obsolete by a publication that appears half-way through your work. Your computer could be stolen, along with the only copies of your dissertation files. Things go wrong. It is a very good idea to allow a few extra weeks in your schedule to cope with all of this. You can do that easily at this stage in your timetabling by artificially bringing forward your deadline. If the real deadline is 1 February, make your deadline 14 January, and aim genuinely to have the work completely finished, bound and ready for submission on that date. This is your lifeboat, your fire escape and your nailfile-in-the-cake. In case of emergency, and only in case of emergency, break into the last two weeks.

7   Add your 'Revised deadline' a couple of weeks above the institutional deadline and write 'emergencies' in the little gap between the two.

Eric says: If you've done your timetable on paper, use a pair of scissors to cut the last two weeks off the bottom of the page. Keep the strip of paper in a glass case with a roll of sticky-tape. If things go badly wrong, smash open the case and tape the two weeks back into your schedule! If your timetable is on your computer, put the last two weeks onto a CD and keep it in a vault in your Swiss bank, only to be spent on your dissertation as a last resort.

There are a number of other constraints that can be added at this early stage, too. For example, if you plan to leave all the word-processing and assembling of text till the end of the project, make sure that you allow yourself time at the end to do so. Writing up bits of the report as you go along is usually a very good idea so that you don't have a huge typing job waiting at the end of the project; we'll say more about that later (Chapter 11). Remember also that it takes a certain amount of time to get your work bound after it is all typed up. How long you need to allow depends on the requirements of your institution. Simple spiral binding can be done on the spot at many high-street printers, but you should allow more time if there are any more specialised requirements (see Chapter 11 for more information). If your submission deadline is shortly after Christmas, don't forget to allow for the holiday period when the chap from the binders, on holiday in the Bahamas, might be remarkably insensitive to your plight.

8   Add 'Start putting final text together' about a month above your 'Revised deadline'.

As you produce your text, of course, you will want someone to look it over for you and check for the silly mistakes that most of us make when we are typing. In many institutions tutors will be happy to read your entire dissertation in draft form. If yours is one of these institutions, take advantage of it; you can get invaluable help at this stage. If your institution is less generous, then at least take advantage of a friend or relative to read through your work. To get any benefit from this process, you will need to give the reader time to read, and give yourself time to do something about their comments. Book your tutor's time well in advance; if you turn up with 9,000 words of text in the middle of a busy week, you might have a long wait for any feedback. Allow at least a couple of weeks for anyone to look at your draft and find time to give you feedback.

9   Add 'Give draft of dissertation to Mum/Fred/Dr X for final comments' about two weeks above 'Give dissertation to binder'.

That gets most of the practicalities out of the way. and your timetable at this stage might look something like Figure 3.3. The space that you have left between 'Now' and 'Final draft manuscript to Tutor' is the time that you have to actually carry out your research and draft the report. As your timetable develops, so the time that you seem to have available to you shrinks! Into your remaining space you now need to insert the major sections of your research programme (as in Table 3.1), and the mini-deadlines by which you intend to complete each section.

Table 3.1 includes a rough estimate of the proportions of your time that you should be thinking of allocating to each of the major tasks involved in the dissertation. You will notice that if you have allowed a complete year to do the work, the numbers in the time-allocation column equate to months. If you have more or less than one year to do the work, then you can expand or compress this timescale. Different types of project may require slightly different emphases on different parts of the work; the allocations given here are a good guide for most projects, but you may wish to fiddle around with them to suit your own project. If you are indeed out of circulation (in China with Eric, for example) for a month, then you need to accommodate that into your schedule at this stage.

10 Slot the key elements of your programme, as illustrated, for example, in Table 3.1, into the remaining space in your timetable, avoiding clashes with prior engagements, and meeting the requirements of the institutional schedule. Add specific dates at the end of each key section to serve as mini-deadlines.

Your timetable should now look something like Figure 3.4. You will notice that in this version the intrusion of a field trip, the summer exams and the trip to China have required us to compress the 12 units of time available in Table 3.1 into less than the 12 months that we thought would be available. Here we have lopped some time off the interpretation period. You will have to figure out your own solution if you face this problem. Try to avoid the obvious, but risky, gambit of eating into your emergency time.

Your own individual timetable

Bear in mind, of course, that different types of project will need different time allocations. For example, if you need to administer a postal questionnaire then you might need to allow for a much longer period of data collection, but might be able to do another job (like drafting your introductory chapters) at the same time. If you need to create a large number of detailed computer-drawn maps, then you will need more time for preparing the report at the end. Figures 3.5, 3.6 and 3.7 show a variety of timetables drawn up for different types of project. Every project will have its own unique timetabling requirements, so you can't just use one of the timetables that we've drawn up; *you* need to draw up *your own* timetable.

One problem that you may encounter at this point is that it can be difficult to construct a detailed timetable before you have decided on the details of your research design (which we will discuss in Chapter 5). For now, you might have to produce a provisional timetable, and do the fine-tuning once you have sorted out the details of the project later on.

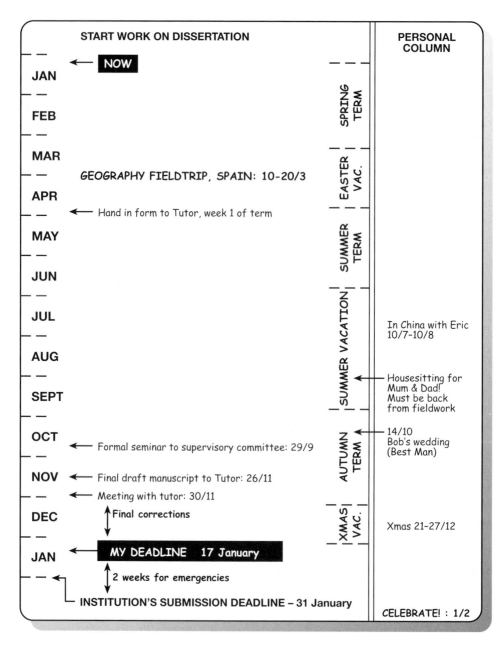

Figure 3.3 What your timetable might look like when you have fitted in all the fixed dates such as terms, administrative deadlines and your safety margins.

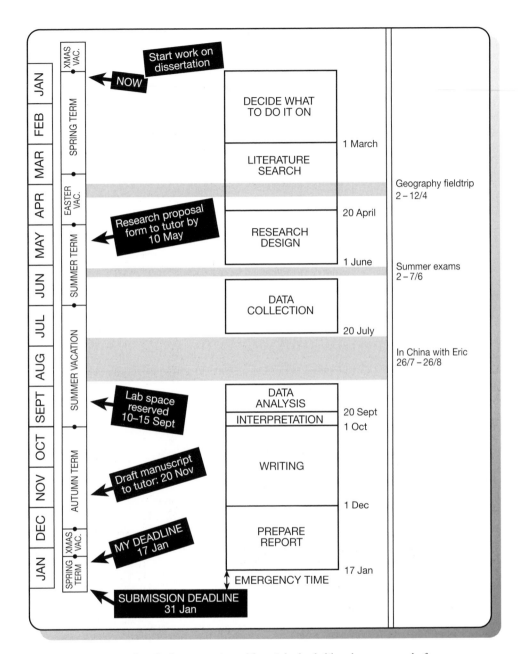

Figure 3.4 An example of what your timetable might look like about a year before your dissertation is due in. More details can be added as you firm up your arrangements later in the year.

Figure 3.5
An example of a
dissertation timetable.

Figure 3.6
Another example of a
dissertation timetable.

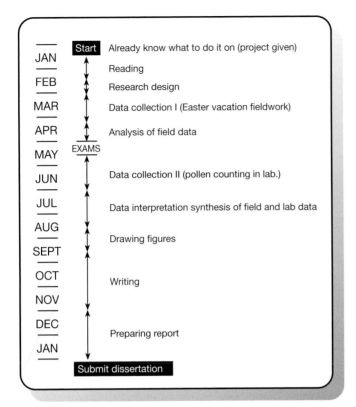

Figure 3.7
Eric's timetable. An over-ambitious programme of data collection leaves too little time for analysing the vast amount of data that might turn up!

Chapter summary and conclusion

It is vital to the success of your dissertation that you allocate sufficient time to each part of the work, and that you do the various bits of the work in the right order. The best way to avoid running out of time or missing out vital bits of the job is to draw up a timetable before you begin. Your timetable should include: all the formal deadlines imposed by your institution; all the constraints on your own time, personal and otherwise; and all the major stages through which you expect your work to progress. Allow plenty of time to deal with unexpected problems. The only way you can produce this timetable is by thinking ahead in great detail about exactly what you need to do, when you will be able to do it and how long it will take.

What to do after reading Chapter 3

When you've finished this chapter, make yourself a timetable!

4

WHAT SHALL I DO IT ON?

This chapter distinguishes topics, problems and questions; discusses your role in selecting a topic, a problem and questions; identifies types of problems to investigate; shows the qualities of good questions; and identifies types of question.

Introduction

In Box 2.1 we listed 'a good problem' as the first quality of a good dissertation. Answering the question that forms the title of this chapter is therefore the first real task you face, and it's by no means a trivial one. In fact, it's *the* most difficult part of your dissertation. Very often, students think that the real work begins after they've chosen their subject. It doesn't. Choosing the right subject, and the one that is right for you, can, and usually does, determine whether you write a good dissertation or a poor one. The aim of this chapter is to demonstrate why the choice of subject is important and what issues you need to think about in choosing a subject. More practically, if, when you've read this chapter, you can complete the sentence that begins 'I am trying to find out . . .' then you will have made the first significant step towards your dissertation.

Topics, problems and questions

Let's begin with some definitions. The *topic* that you do your dissertation on is some broad area of study. It may be retail geography, hillslope geomorphology, rural accessibility, land tenure in eighteenth-century Devon or any such branch of the discipline. The *problem* that you tackle will be a more specific, and smaller, issue within this topic. So, for example, within the field of retail geography you might tackle problems such as the decline of retail services in Northamptonshire villages or the catchment area of an out-of-town shopping mall. Finally, there will be specific *questions* that you will ask in relation to your problem. Have retail services declined at the same rate in all Northamptonshire villages? Has the rate of decline changed through time? In reaching the point at which you can complete the sentence 'I am trying to find out . . .' you will need to select a topic, identify a problem and specify questions.

Selecting a topic

The obvious approach to selecting a topic, and the one which you will most often be advised to take, is to choose something that interests you. This is a sound approach, but not always the best one. First, you may be most interested in things that do not easily lend themselves to a dissertation study. Some things are inherently difficult, expensive or time-consuming to study. Your resources of intellect, finance and time will be limited. You are advised to sort out what these are at an early stage. Make a list – some issues that you need to consider are listed in Box 4.1. Second, your institution may have rules about the need to collect *primary* data. This may be taken to mean that you cannot do a study of housing in Harare unless you actually visit the place. In other institutions, however, it might be quite acceptable to use data that somebody else has collected (including census data) provided you make due acknowledgement of this fact. Check the rules in your own department. Third, there may be many things that interest you, so how do you choose which one to study? If you are in this fortunate position then you might pursue several possibilities simultaneously. In all likelihood, one of them will soon turn out to be more promising and interesting than the others. On the other hand, you may not find that anything at all interests you very much! You may get quite an inferiority complex if you see your friends have topics (and therefore, you think, have a part of the discipline in which they *are* interested) and you haven't. So, what other approaches are there?

Box 4.1 Things to consider before selecting your topic

- Is the location fixed or limited to a few choices?
- How much time do you have for data collection?
- Can you work on the project full-time for a period or do you need to do something else for part of the time?
- Do you have people who could act as field assistants? If so, how many?
- What transport do you have?
- Do you have money to spend on this work?
- Do you have access to unusual data sources?

Try turning the problem around. Think about the things outside of geography and your lecture courses that *do* interest you and see if there is a way in which you could bend the geography to match your interests. You may be a great rock climber. Few geomorphologists are; so there is not a great deal of work done on the geomorphology of vertical rock faces. Consequently, this is an area in which it is relatively easy to come up with a worthwhile problem (but there may be safety issues that you will need to consider). When you think about it, there is a geographical aspect to most things so it shouldn't be too difficult to find a topic in even apparently unlikely fields. It's a mistake to separate your geography from the world you live in. The danger is that you may spend more time on the other interest and less on the geography. However, if you don't fall into this trap you could end up with a good dissertation *and* a greater appreciation of some aspect of your leisure interest. As an example of this approach, you might look at *Scotland's Golf Courses* by R. Price (1989). The golf courses of Scotland have some very interesting geomorphology! If you know that you will be spending part of your summer in China with Eric, do you have an opportunity to undertake your dissertation there also? If so, try looking at the literature on the area to see if a topic suggests itself.

You can always try asking somebody's advice. To do this before you've come up with some ideas of your own is a bit wimpish! You've got a year's work to do and you can't even get past step one! Nevertheless, getting started is always the most difficult stage for most things. Who knows, one small piece of advice at this stage could set you off towards a prize-winning (Chapter 2) effort. But a word of warning: don't go into your tutor's office and start with whining 'What can I do my dissertation on?' This will not go down well. Try seeking out, say, the transport specialist in your department and starting off with: 'I've been thinking about studying the railway network of France for my dissertation'. This may be surprisingly productive. On the other hand, it may not; your department may have strong views on your role and that of your supervisor in choosing what you do your dissertation on.

In many departments choosing the topic of your dissertation is regarded as part of the training in research, and your choice of topic is something that the examiners will judge you on.

Student choice and off-the-shelf projects

You may have thought that student choice was an expression that always worked to your advantage. When you have to choose a dissertation topic for yourself, however, the advantages may be less than obvious.

Eric says: Can't somebody else choose my dissertation for me?

Although some departments are prepared to identify a list of projects for students to choose from, most are not and expect students to come up with their own dissertation projects. There are good reasons for this. Deciding what to investigate is a very important step in research. If your tutors did this for you, it would deprive you of an important part of your research training and experience. In leaving this step to you it is expected that you will learn what constitutes a good topic and what does not, and also that you will learn both how difficult and how important it can be to choose your topic wisely. Consequently, when you have your first conversation with a potentially useful tutor in your department, you might be disappointed to discover that the tutor doesn't simply give you a straightforward suggestion about a good topic, but may start asking you difficult questions about the topic. The aim of this is to make you think more carefully about your chosen topic, and may be to make you realize that this is not a good topic for you to work on. Don't be disheartened if this happens, and happens more than once. If, at the end of this stage, you not only have a good dissertation project but also know the difference between a good project and a bad one, all the hard work will have been worth it and you will have achieved one of the goals of the dissertation project.

In the same way that your tutors will probably not give you a ready-made project, they will probably not encourage you to adopt ready-made projects from other sources. For example, several commercial and scientific organizations offer you the opportunity to combine your dissertation research with an organized expedition as part of their operation. While these organized expeditions do provide a good opportunity to travel to locations that might be difficult for you to reach independently, and therefore open up a whole new range of opportunities for your dissertation, there are also some problems with following this route. The first problem is simply that your institution may not allow it. Check your regulations and talk to your tutor. The more pre-arranged your dissertation seems to be, the less likely your tutor is to allow it. The second problem is that if you choose a topic from a list provided by an external organization, your examiners won't be able to give you any credit for devising a project of your own. A third problem with some organizations is that they offer field support and even project supervision that will effectively turn you into a research assistant: dependent upon the operation for every detail of your fieldwork, and possibly receiving direction that conflicts with the requirements of your dissertation. We don't mean to sound too negative about organized expeditions because some of them can be very effective and can help you a lot. However, we do mean to warn you to be very careful before paying money and committing your time to an expedition that may not, in the end, provide you with what you need for your dissertation.

Eric says: **Maybe my friends and I can organize our own expedition?**

Identifying a problem

So, you think you've got a topic. Next, you need to identify the specific problem that you will work on.

Now is the time to do some serious research. Explore the literature.

Identifying a specific problem involves pretty much the same thing as choosing a topic all over again, except that it's on a smaller scale. Say you've decided to do your dissertation on rivers in your local area. Just what is it about these rivers you are going to investigate? The approach you will need to take to this stage is likely to be somewhat different. To select your topic the emphasis was on what you are interested in and are capable of undertaking. To identify a problem you need to become familiar, if you are not so already, with what other people have done and have found interesting. Your dissertation needs to fit into and grow out of the context of the existing literature on the topic. Now is the time to explore that literature in detail. Visit the library, either in person or on-line and start looking through the literature. You need to find the specific research problem you are going to work on, and you also need to find out what work has already been done on this problem.

To find a worthwhile problem within the topic area you have selected, you should explore the literature using several different approaches. You can look through the shelves of books and journals in the library. You can search bibliographic data-bases such as GEOBASE. You can search web-based search engines designed for your particular field of research (such as SCIRUS for science-related subjects such as physical geography).

First of all, you should check the most recent literature in order to find out what topics are currently being discussed most heatedly; in other words which are most topical within this subject. Once you have identified a 'live' topic, you should delve further back into the literature to trace its roots.

Start by going through recent (say, the last five years') issues of the journals that you know cover your topic. Journals that have review articles or progress reports on particular fields of research can be particularly useful at this stage. Both *Progress in Physical Geography* and *Progress in Human Geography* regularly carry review articles and progress reports. Another way in which you can quickly gain an idea of what is being done in a particular field is to search an on-line database such as GEOBASE, which will probably be accessible through your institution's library or web pages. The short abstracts that these databases provide may be enough to tell you that, even though the title looks promising, the article isn't relevant. Alternatively, they may tell you that this is something you should read, and is worth downloading, getting out of the library, or obtaining on inter-library loan if your library doesn't have the journal. These databases will list large numbers of papers relevant to your search terms, and they usually provide at least an abstract for each paper and in many cases links to copies of the whole paper. The trick is to enter appropriate search terms that will call up appropriate references from the database. Abstracts are short and easy to scan quickly for relevant details, so reading through many pages of article titles and reading a few abstracts of articles that look interesting to you won't take a great deal of time.

Of course, you may be lucky and have a journal almost wholly devoted to your topic area. For example, if you have decided to work on periglacial geomorphology then the journal *Permafrost and Periglacial Processes* is a must. Don't stop there, however. Not all the relevant, or even the best, research is in the specialist journals. Researchers working in a particular field will always want to communicate their findings to fellow workers (the readers of the specialist journals) but sometimes they will also want to reach a wider audience, especially if they have something to report that has wider significance. The journal *Nature* doesn't have many articles on periglacial geomorphology, but any that it does have will probably be very important. So, although going through the last five years of this weekly publication is a lot of work with a very slim chance of a reward, it is certainly worth using the index to check whether any periglacial papers appear. You can also tailor your database search to look only in specific journals. This will help to keep the number of references returned within manageable proportions! Reading the recent papers on your topic, or even just their abstracts, will soon give you an idea of the sorts of problems that are being investigated and the methods that are being used. You will probably be planning to do a web search using appropriate keywords. While this is almost certain to lead to some material, it is also the least reliable, because, unlike the contents of journals, what is on the web is unlikely to have been put through the review process. Whereas what you read in journals has been checked by people other than the author before it is published and these people have stated that what is in the article is worth publishing, this is seldom the case for material on the web.

Remember, any old idiot can put up any old rubbish on a web page.

For this stage of your dissertation, try to find one or two days when you have no other commitments (lectures, tutorials, etc.) and spend the entire time searching the literature. Give yourself the opportunity to get to know your topic. Don't be in too much of a hurry to identify your problem. You may think you are spending a lot of time and not getting anywhere, but this will be time well spent. What's more, if you make notes of the things you've read you will find these notes useful when you come to write up the intellectual context part of your dissertation. This is probably the most important period of work on your dissertation, so take your time and do it properly. If you come up with a good problem, you will be well on the way to being able to produce a good dissertation. If you don't, you won't.

During this stage of your project you will start to accumulate a collection of references to previously published research. In due course you will use some of these in your final write up, so you might as well collect and file them efficiently from the start. A lot of students waste a lot of time searching for the details of references that they have misfiled or lost, and it can be a frustrating way to delay your completion! At the every least you should write down the complete reference for every item that you use in putting together your project. You may wish to use one of the many bibliographic software packages that are available to manage your reference collection. These not only allow you to store and annotate references electronically, but also interface with your word-processing software to allow you to produce reference lists very quickly after you have created your text. Some students find such software useful, but be careful not to waste time on any software unless you really find it helpful.

At the end of this stage you should be able to state the problem you are going to work on. From your reading you may have come to the conclusion that there are aspects of your chosen topic that are ignored in the literature and that you could undertake a dissertation that would address one of them. For example, in looking through the recent literature on rivers you may conclude that there has been little study of the channel geometry of rivers flowing in poorly sorted glacial deposits. It so happens that you live in an area full of such rivers and are thus ideally placed to look at this aspect of the topic. You have your problem (or, at least, you may have). There are a number of ways in which you can arrive at a problem to investigate from reading the literature; some of these are listed in Box 4.2.

Box 4.2 Some ways in which you might identify a research problem from reading the literature

It appears that nobody has investigated this topic – I'll have a go.

Parsons and Knight (1998) investigated this topic and raised the question regarding the role of x. I'll investigate the role of x.

Parsons and Knight (1998) investigated this topic and found out . . . but they ignored the possible effects of x. I'll investigate the effects of x.

Parsons and Knight (1998) investigated this topic at location x and found out that . . . I'll see if the same is true for location y.

Parsons and Knight (1998) investigated this topic and found that . . . I wonder if things have changed since then. I'll repeat their study and compare my results with theirs.

Parsons and Knight (1998) investigated this topic by method A. I wonder if you get different results by using method B. I'll use method B and compare my results with theirs.

Since Parsons and Knight (1998) did their study a new data set has become available. I wonder if the new data set supports Parsons' and Knight's conclusions.

The next step is to find out whether somebody else has already tackled this problem. If they have, it doesn't mean you can't still work on it – after all, you may completely disagree with their findings. However, it is necessary to know what work has already been done. There are three approaches, and you should consider trying all of them.

The first is to look at published abstracts and bibliographic databases. These provide a quicker route to finding out what has been written than reading back-issues of all the relevant journals. They are particularly useful if you know exactly what you are looking for.

The second approach is to talk to your tutor. Now that you know what your problem is, and have read some of the background literature, you will be in a position to have a worthwhile conversation. Your tutor will take you more seriously as a result of your evident knowledge, and you will gain far more from what he/she has to say because of it. It's quite likely that your tutor will have relevant journals that are not in the library (this will be faster than waiting for inter-library loans) and may have been to conferences and heard papers that have yet to be abstracted (the abstracts are usually up to two years old) or even published. In addition, your tutor may know relevant experts in the field who could offer useful advice. These experts offer the third approach to finding out what has already been done.

You can email, write to or visit these experts. But be warned: experts in a field are unlikely to be very pleased if they have to deal with ill-informed students with poorly focused problems. If you approach them in this state, not only are they likely to be less helpful than they might be but you are in danger of making them suspicious of the next student who approaches them. It might be a good idea to check with your tutor before you start writing to strangers who might get upset at being pestered by somebody else's students. Receiving an email from some undergraduate in a distant location about their dissertation is not something people, in general, look forward to!

Don't contact an expert in a field until you have a fair degree of expertise yourself.

To get anything worthwhile it is important that you appear serious about the topic and that you are approaching your expert not because you think he/she will offer a short cut to a good dissertation, but because you are already on the way to a good one and are looking to make it outstanding!

Finally, don't be too disheartened if, even at this stage, you discover that your problem cannot be investigated within the limits of your dissertation, or even that the topic is not suitable. As we said at the beginning of this chapter, what you do at this stage will play a large part in determining how good your dissertation is. We'd all like our research to progress smoothly, but it seldom does. Difficulties at this point, however, are particularly annoying. You want to get going and you can't even sort out the direction to go in! But it is better to spend time thinking about it now than set off in the wrong direction and have to turn back much later.

This bit of research is probably the most difficult. Don't feel you are stupid if you seem to be having problems with it. It's almost certain that you will.

Specifying questions

Specifying questions is all about *how* you approach the problem you have decided to investigate. The ability to specify the right questions is what distinguishes a good researcher from a poor one. Take the example of Richard Feynman (whom we've mentioned before) who, while he was working at Los Alamos building atomic bombs, found he had spare time and so decided to investigate the problem of cracking safes (Feynman 1985). The safes he wanted to crack had combination locks and to open them required a sequence of three numbers, each in the range 0–99.

One way of cracking the safes would be to try all possible numbers (1,000,000). Feynman realized that this would be unreasonable. He had the insight to specify the question 'How accurate does each of the three numbers have to be?' If the correct combination was 23–66–49, would 24–67–48 work? If you could be quite inaccurate and the safe would still open then trying all the necessary numbers would become a reasonable approach. If being within five either side of the correct number would do, then trying all the necessary numbers (0–0–0, 0–0–10, 0–0–20 and so on) wouldn't take very long because there were only 1,000 of them. Feynman discovered that for the safes used at Los Alamos he needed to try only one number in five, and was on his way to becoming a renowned safecracker. Likewise, it has been said that when the apple fell, Newton could have asked the question, 'Why did that apple fall just then?' and been on his way to not very much at all. His genius lay in specifying the question, 'Why do apples fall?'.

Specifying the right questions

So, how do you know what are the right questions? By their answers shall ye know them!

Only bother with questions that look as though they will have interesting answers.

The right questions are ones that have useful answers. These questions lead to new knowledge, or solve a specific problem. In the case of Feynman, if the answer had been that the number had to be spot on, he would have known that if he wanted to be a safecracker, he'd have to find another way of doing it. To discover if this was the case, all he had to do was to try a safe for which he knew the right combination and see if he could be wrong by one digit, two, and so on. As it turned out, he discovered that the problem was tractable using the approach he had thought of. The question was easy to answer, had two possible answers, and each led to a significant advance in the problem he was investigating. The problem was how to crack safes. The question was, 'Can you do it trying all the numbers?' That led him to a further question, 'What is the size of the set of all the necessary numbers to crack the safe by trying all these numbers?' If Feynman had discovered that the set had 1,000,000 numbers in it, he'd have known that he'd have to try another method. If the set of numbers was sufficiently small, then he knew he had a workable method. So, not only had he specified a question but he also knew what possible answers the question had and what each of the answers would mean in terms of the problem he was investigating. Questions are the right questions if you know what answers they may have and what these answers will mean to you. In addition, questions that are easy to answer, and by answering them you make a significant step forward in your investigation, are even better.

Types of question

Because questions are so fundamental to good research, it is worthwhile looking into their nature a bit further. In this section we want to show that, although there may be an infinite number of questions you could ask, there are only a few *types* of question. By and large, how you can tackle a question depends on the type of question it is. Consider the questions in Box 4.3. Questions such as these might form part of the problem you are investigating for your dissertation. Although these questions may all appear to be different, in effect they are all of the same type. They are all asking for a description. A very large amount of geographical research can be put into the category of this type of question. We might generalize the question to the form 'What is *A*?' or, even more succinctly. '*A* =?' For example, Question 3 in Box 4.3 can be rephrased as 'What is the number of visitors who use this recreational centre and what are the times at which they use it?'

Box 4.3 Various questions that are all of the same type

1 What is the catchment area of the Potteries Shopping Centre?
2 What is the vegetational history recorded by this peat core?
3 How many visitors use this recreational centre and when do they use it?
4 What is the attitude of long-term residents to recent migrants to this town?
5 What proportion of fields in this area shows evidence of erosion by running water?

In Box 4.4 we list many other questions that, similarly, could form part of your dissertation. We show a way of grouping these questions into types, and we give all the types of questions that we have identified similar, succinct generalized forms. How you approach answering a question depends, to a degree, on its type. We will refer to our types of question frequently in the following chapters.

Types of answer

Different types of question demand different types of answer. This is useful in several ways. For example, if you are having trouble coming up with a good question, try imagining what you would like to say in your conclusion to your dissertation, assume that to be your answer, and work backwards from there to work out what question you need to ask in order to head towards that answer. The fundamental link between question type and answer type is also useful because it means that as soon as you have worked out what type of question you are asking, you will also know what your answer will have to look like. If your question asks 'Why?', for example, then your answer will have to say 'because'. If your question asks 'What causes . . . ?', then your answer will have to say 'this does'. This means that you can check constantly, as your project progresses, whether your research is really leading you in the direction of an appropriate answer to your question. Only if you

Box 4.4 **Our types of question**

Questions	Generalized form of question	Succinct generalized form of question
What is the catchment area of the Potteries Shopping Centre?	What's A like?	A=?
What proportion of fields in this area show evidence of erosion by running water?	ditto	ditto
Do these two deposits of glacial till have the same provenance?	Is A like B?	A=B?
Are the children in all the schools in this town drawn from the same range of economic backgrounds?	ditto	ditto
Do these two deposits of glacial till have a different provenance?	Is A different from B?	A≠B?
Do the schools in this town differ in terms of the socio-economic backgrounds of their pupils?	ditto	ditto
Is it better to measure grike width rather than grike depth to characterize their age?	Is A better than B	A>B?
Does the record of annual expenditure on agricultural lime provide a better measure of rural wealth than the record of grain sales in eighteenth-century Shropshire?	ditto	ditto
Is there a relationship between the number of respiratory deaths in a town and the size of the town?	Are A and B related?	A↔B?
Does the distribution of oak in Britain correlate with the distribution of holly?	ditto	ditto
Does mean river channel width increase as discharge variability increases?	Does A affect B?	A→B?
Does the amount of usage of public transport increase as the frequency of the service increases?	ditto	ditto
Does increasing the frequency of a bus service cause more people to use the service?	Does A cause B?	A=>B?
Does moving to a larger town increase your chances of dying from a respiratory disease?	ditto	ditto

know what your answer is supposed to look like will you be able to judge whether your answer might be fundamentally flawed when you eventually reach it. Making sure that you arrive at an answer that matches the question you asked is one of the jobs of research design, which we will consider in the next chapter. In the meantime, look back at the different types of question that we identified in Box 4.4 and see whether you can work out what would make an appropriate type of answer to each one. And what wouldn't.

Chapter summary and conclusion

This chapter has looked at the difference between a topic, a problem and a question; identified the qualities of good questions; and shown how, although there are many questions, there are only a few types of question and that they can be represented by succinct generalized forms.

What to do after reading Chapter 4

1 Name your topic.
2 Define your problem by writing a sentence that begins 'I am trying to find out . . .'.
3 Make a table with three columns. In the first column, list the questions that you will ask in order to solve your problem. In the second column, list the possible answers that you can get to each of your questions. In the third column, state what you will know if the answers to the questions turn out to be those listed.
4 Explain to your Mum what you are going to do your dissertation on.

5

HOW DO I START?

RESEARCH DESIGN

This chapter discusses different ways of getting answers to questions, your role and that of a theoretical framework in answering them, research design logic and research designs for different types of project.

Introduction

By the time you get to this stage, you should know what it is you are trying to do. You should have chosen a topic, identified a problem, specified the question(s) that you will tackle, and know the type of answer(s) you are looking for. This can be a very difficult moment.

You've done all the planning you can think of and now you have to *do* the dissertation. It might be difficult to know where to start and the task ahead might suddenly look very big. Lots of people get this feeling. It's a sort of mental barrier like writing the first word on a clean sheet of paper – not writer's block but doer's block. This is one of the reasons why a lot of students keep putting off the dissertation longer and longer and eventually have to do it all in a rush at the last minute. Don't be like that. You should have left space in your timetable to settle down and sort out in your mind exactly what you have to do. When you can see the steps ahead, progress becomes much easier. So, you know what your question is; now you want to know how to get the answer. That's what research design is all about.

Getting answers to questions

The aim of this chapter is to alert you to the issues you will need to consider to obtain the answer(s) to your research question(s). We can't tell you how to do your particular research project. What we can, and will, do is discuss the issues and give some examples. The examples will, we believe, contain something useful for you. You should take from them those parts that are useful.

There are lots of ways of getting answers to questions. For example, if you wanted to know how to get to the railway station, you could just ask someone. They would give you an answer, but of course you wouldn't know how much faith to put in that answer. If you wanted an answer that you could be more sure of, you could ask a policeman, look in the A-Z, or phone the railway to ask. If your different sources gave different answers, or to convince yourself thoroughly, you might have to go over to the station and look for yourself. The point is that there are lots of ways of getting answers to any question, but sometimes the different approaches give different answers, and it isn't always easy to judge the reliability of the answers you get. Also, it is not necessarily the case that one answer is right and all the others are wrong. One person you asked about the way to the railway station might have assumed you were walking, but another may have assumed you were driving. Both answers may have been right (or wrong!) depending on your perspective. When you do your own research project, you need to find a way of getting answers that enable you to judge their reliability from a particular perspective. You need to be able to reach conclusions with confidence. It's no good getting an answer to your question if you don't know whether it's the right answer or not. You need to design a research programme that will lead you to a conclusion that is reliable in the context in which your research is set.

The role of theory

Human geographers, and social scientists in general, often talk about 'theory' very explicitly in their research, but physical geographers and other physical scientists tend not to. Why is this, and how does this difference affect your dissertation?

Eric says: I am a physical geographer. I eschew my theoretical foundation!

All research is underpinned by theory; that is, it is set in a context of previous ideas. That's what enables us to go forward and find out new things. So, if your dissertation is concerned with sediment transport on hillslopes, it is set in the context of Newton's theory of gravity. Your examiners won't, however, expect you to include a discussion of this theory in your dissertation (you may be pleased to know!) because they know that everybody accepts this theoretical foundation for research on sediment transport. Quite simply, we have one, generally accepted theoretical foundation for research on sediment transport, so there's nothing to discuss (at least, not about Newton's theory of gravity in an undergraduate dissertation in geography). But if your dissertation is about gender perceptions of New Urbanism, your examiners will expect a discussion of its theoretical foundation. This is so because there are several theoretical foundations you might adopt to answer your question. You might ask that question from an explicitly feminist perspective, or from a Marxist one, and your theoretical standpoint would determine your approach. There are fewer opposing theoretical standpoints in physical geography. Going back to our analogy about asking the way to the railway station, it's as if all physical geographers had decided a while ago that the appropriate way to get there was by bus, so they were looking for an answer about bus stops and timetables, but the human geographers had a variety of different ideas about how to travel: some want to walk (so were collecting information on footpaths and pedestrian crossings), others to drive their cars (so they wanted to know about speed limits and one-way streets), and yet others to go on horseback (and they were looking for bridleways and horse troughs). That's a bit of an oversimplification, and there are theoretical issues about how questions in physical geography are tackled (see Rhoads and Thorn 1996), but the majority of research in physical geography does not address them. In your dissertation, it is important to decide early whether you need to address theoretical issues in your research or whether the theoretical underpinnings of your research topic are sufficiently uncontroversial that you can take them as given.

Types of question

The research design that you employ, in other words the way you go about your dissertation, depends on what you are trying to find out. That sounds like common sense, but failure to bear it in mind has led to the downfall of many promising projects.

There may seem to be an almost infinite number of different research projects for which you might be trying to design a research programme. You might be trying to find out when a medieval settlement was abandoned, why a river has changed its course, whether an influx of refugees has had an impact on a local economy, where best to locate a new cinema, or how to reduce beach erosion. You could be trying to find out almost anything when you do your dissertation. However, you can reduce this almost infinite list to a very short list if you think not about individual titles, but about what *type* of question you are asking. Are you asking a when question, a what question, a why question, or a how question? Is it a who, or a does, or an is, or an are, or a will or a should? Different types of question demand different types of answer. For example, a 'why' question demands a 'because' answer. A 'how' question demands 'like this'. 'When?' demands 'then'; 'is?' demands 'yes' or 'no' (or 'sometimes'); and so on. You can therefore define your question in terms of the kind of answer you are looking for.

Are you looking for a 'yes', a '27' or a 'because the ground was wet'? Do you want a date or a location? Are you looking for a process? Are you looking for the murderer, the murder weapon, the motive or a judgement of guilt?

One of the most commonly used distinctions is among studies concerned with description or narrative (what is the world like?), studies concerned with explanation (why is the world like it is) and studies that form judgements (should the world be like it is?). Many historical studies, in both human and physical geography, are descriptive. For example: what was the vegetation like here at this point in history? When did agriculture begin at this place? What has been the history of climate change at this place? Many regional studies tend towards narrative description. For example: 'Landscapes of East Anglia' or 'Tin mining in Cornwall'. Explanatory studies are different from descriptive studies in that they tend to look for causal relationships, and therefore are often process-related. For example: Why was vegetation like this at that time? What caused agriculture to begin at that time? Why does crime have a geographical dimension? Does increased volcanism cause climate to change? There are many philosophical arguments about what constitutes explanation, and this question has led to much discussion in the social sciences (see, for example, Goodman 1978; Ryan 1970) about appropriate ways to identify causal mechanisms for social phenomena, or even if causal mechanisms can be identified. If your dissertation question is one that questions why something happens or exists then you need to give some thought to what *you* mean by explanation. At one level, you might argue that an explanatory study is no more than a sophisticated description (you are describing the relationship between two phenomena, for example volcanoes and climate).

The third type of question, about judgements of value, is one that physical scientists would not generally accept as falling within the realm of scientific research, but one that social scientists would consider as a valid research topic. A physical geographer is more likely to reach a conclusion that *if you wanted to change the environment in this way*, this research provides a method for doing so. A human geographer is more likely to conclude that we should, or should not, wish to change the environment *to make the world a better place*. The physical geographer might feel that deciding what was better fell outside the scope of the work. The human geographer might see that as fundamental to the work.

At the end of the last chapter you wrote a sentence beginning 'I am trying to find out . . .'. Make sure now that you are clear what kind of question you are asking, and what kind of answer it demands. As we said in Chapter 4, it is sometimes helpful to break your question down into an almost algebraic form. Replace all the nouns with letters, leaving only the verbs or interrogatives to define your project. For example, if your project is 'Does forest clearance in Grampian Region cause accelerated soil erosion?' it will become 'Does A cause B?'. Reduce the title still further by the use of symbols to replace the interrogatives and verbs. Thus 'Does A cause B?' becomes 'A$\Rightarrow$B?' If you try this with a number of different project titles, you'll begin to see that all the different projects you can think of fall into only a small number of categories. There are only a small number of types of question. This simplifies the problem of finding the right research design for your project.

Where do *you* fit into doing your dissertation?

Throughout this book, we have emphasized your role in your dissertation – the benefits to you in doing it, how you can use your own interests to select a topic and so on. What role do you, as an individual, have in the doing of it? Would the outcome be just the same if you handed the whole thing over to Eric at this point? There are three possible answers to this question:

1   It should be just the same if Eric did it, and you should strive to make it so.
2   It wouldn't be the same, and we must reluctantly acknowledge this fact.
3   It shouldn't be the same if Eric did it.

Most physical scientists would think the answer should be (1). Whereas you have an important role in deciding what question to ask, how the question is answered should not depend on your identity. This answer underpins the notion of a repeatable scientific experiment. If you were to drop balls from the leaning tower of Pisa you should get exactly the same result that Galileo did. However, research isn't quite as straightforward as repeating somebody else's experiment, particularly in a field-based subject. No matter how carefully you try to design your research, and no matter how far you think you have a research design that will cover all eventualities, you can be sure that the unexpected will arise. You are likely to find that the method for collecting your soil samples, for example, has failed to provide a procedure for deciding what to do if your sampling point lands on a fallen tree trunk. Inevitably, while actually doing your dissertation, situations will arise where decisions about procedure have to be made. It's at this point that Eric might make a different decision from yours. (You decide to move the fallen tree trunk, but Eric sets up a rule for choosing a new sampling point when the original one was inaccessible.) If they are well thought out in the context of the research question, both your decision and Eric's are equally valid. Provided you document how you dealt with such matters, you are well on the way to having a repeatable experiment. However, just as you may have found unexpected occurrences when you tried to collect your data, so somebody else trying to repeat your work could come across an eventuality that you hadn't encountered and for which you therefore hadn't documented a procedure. So, even if physical scientists might like the answer to be (1), there are good reasons why it might turn out to be (2).

Many social scientists, on the other hand, think the answer should be (3). If you are conducting a piece of research that involves your interaction with a community as part of your study of how that community works (for example, conducting interviews), then your identity and the way you interact with your subject will influence both the outcome of the research and the way you subsequently write up your project. If Erica, rather than Eric, were conducting the interviews, she may obtain different responses from those Eric would obtain (this issue is sometimes referred to as 'positionality'). Working with different material, even if both Erica and Eric used the same approaches to their analysis, they may well reach quite different conclusions. Some social scientists would argue that because your identity is so closely linked to your research, you should involve yourself wholly in writing it up. Others, however, would argue that there is an important distinction between a recognition of the subjectivity of research findings and a personal story told by Eric that may be of no interest to Erica. Either way, you need to give some thought to your identity and its implications for your research design. Can *you*, in fact, obtain the answer to your research question? Making sure that you can reach a reliable answer, from the starting point of your question, via all the obstacles or opportunities presented by your identity, your theoretical background and other constraints is what research design is all about.

Research design

Research design is critical to any piece of research. It is like the architect's drawings for the builders; it is like the Field Marshal's battle plan; it is like the explorer's map for the treasure hunter. Without a sound research design your research will get lost, your project will be defeated and your dissertation, if it ever gets built, will fall down. It is hardly surprising, then, that scientists and philosophers have dedicated much effort to the study of research design. However, although there is substantial literature into which you might delve, no consensus has really emerged as to a universally applicable 'best method' or 'best way of finding out'. Partly because there is such a choice of approaches, and partly because not all researchers have taken as much time as they should to consider those choices, you will be able to find all sorts of different types of research design if you look in the current research journals in your field. You should be able to judge for yourself that some are more effective than others, and that different approaches are valid for different types of research. As we said earlier, there are lots of different ways of getting answers to questions. Your task is to find the best one for your project.

It is very important that your research design is sound, and although all this 'philosophical' background might seem far removed from the business of your dissertation, it is, in fact, central to it. It is worth being familiar with the different sorts of reasoning that you might be using, so that you are in a position to make sure of using them properly and not laying yourself open to a painful grilling from the examiners! We said in Chapter 2 that in your dissertation you have to find out something. If you reach the conclusion in your dissertation (as we hope you will) that you have found out X, then you have made a claim to knowledge. Philosophy is all about the validity of claims to knowledge (sometimes referred to as epistemology). You need to know (and demonstrate to your examiners) that you are aware of the perspective of your claim and the assumptions on which it rests. If everybody was agreed that there was a 'best' way of finding things out and that there was only one perspective from which knowledge could be regarded as valid, you wouldn't have to worry about this in your dissertation. If you ask somebody in the street what time it is, it is assumed by both of you that you mean local time, and not the time in Hong Kong, so you don't need to discuss the issue. But, as we illustrated earlier, if you ask the way to the railway station, there are different perspectives from which you could be asking this question, and you do need to make your perspective clear to the person you are asking.

Examiners' favourite questions for a viva include:

1 Do you really think that the critical rationalist approach you have used is appropriate for this sort of project?
2 Could you explain the logic of your research design?
3 Did you consider a qualitative approach to data collection?
4 Can you explain what you were trying to do?

There is a lot of literature on approaches to research and you are almost certain to have come across some of it in your course, particularly so in social sciences where there is less of an agreed 'best' approach than there is in the natural sciences. We don't intend to discuss all of these approaches here, because that is not what this book is about. What we will do is look at a few questions and the implications of taking different approaches to address them. The aim is to illustrate how you should now think about your research question(s), and to identify which approach is best for you and for the question(s).

Dissertation question 1: Does social group affect tourist behaviour? (algebraic form of dissertation question 1 is A→B?)

You might tackle this topic from a variety of approaches. If you took a behavioural-ist approach you might seek to understand how tourist behaviour varied among social groups according to their response to information about the tourist destination that was presented to them. If, on the other hand, you took a Marxist approach to the problem, you might seek to show how inequality of use of the tourist destination resulted from economic and social deprivation of some social groups. But, equally, you might take a feminist approach in which you sought to show how differences in oppor-tunities provided by a tourist destination arose from a gender bias in the way those opportunities were presented to tourists from different social groups. All of these approaches would lead to insights into the question posed as the dissertation topic. But they would lead to different insights. The quality of the insight would depend both on how well you had conducted your research *and* on the usefulness of the approach.

Dissertation question 2: What is the effect of substrate quality on the decomposition dynamics of plant residue? (algebraic form of dissertation question 2 is also A→B?)

If you took a critical rationalist approach to this question you would set up hypotheses and seek to falsify them. (For example, that a less permeable substrate accelerates the decomposition dynamics of plant residue.) Setting up several such hypotheses may not be an efficient way of answering the question. On the other hand, if you took an inferential approach you would explore the effects without any precon-ceived hypotheses. This would be a useful approach if you already knew (or were prepared to assume) that the substrate had an effect and you were seeking to quan-tify it. Such an approach might be a precursor to a subsequent study using your conclusions as its hypothesis.

Dissertation question 3: What is the role of agroforestry in the UK? (algebraic form of dissertation question 3 is A=?)

Here you might take a structuralist approach in which you focused upon economic and social frameworks that could influence the role of agroforestry. But you might also take a behaviouralist approach where your focus would be farmers' attitudes to agroforestry and how those attitudes influenced the role of agroforestry. Again, these approaches would lead to different insights into the question posed as the disserta-tion topic.

From the range of conflicting ideas about research design that have been put forward, we can draw out some key concepts that will help you to sort out your own research and, in particular, how you conduct your research within a particular frame-work. These are the concepts of logic, which need to be applied to any type of research design, and of induction (or inductive reasoning) and deduction (deduc-tive reasoning), which in themselves represent specific types of logic or specific types of research design.

Logic

Logic is the branch of philosophy concerned with reasoning. From our point of view logic is all about distinguishing good arguments from bad ones, and about ensuring that our scientific reasoning is sound.

Your dissertation, or any piece of research, is an exercise in reasoning, so logic is very important in your research design.

From our everyday experience we are all familiar with 'working things out'. For example, we might have to work out what time to set the alarm clock in order to make a morning lecture, allowing time for breakfast, for walking the dog (unless Eric can come over and walk him for us) and for the vagaries of the bus service. In that situation we would check the bus timetable, ask Eric if he could come over, allow the usual time for breakfast and from this, work out the time to set the alarm. Our powers of reasoning enable us to do that, and our skill in logic enables us to reason soundly. If Eric could walk the dog, we would be able to set the alarm later. If we were late for the lecture and said it was because we had to walk the dog and the alarm didn't go off, our punctuality could be criticized but our logic would be impeccable. If we were late and said it was because Eric walked the dog so we had more time, no one would be able to follow our reasoning. It would be illogical. (NB: this can be a useful technique when trying to excuse your late essay etc.!) With everyday examples it is usually easy to spot 'bad reasoning' or illogical argument. In research, where the subject matter is less familiar and the 'right' answer is less obvious, it isn't always so easy. That's why we always need to be sure that our logic is in order and our reasoning is sound. If they are not, then the conclusions we reach will be invalid (wrong, meaningless and worthless).

In philosophy, logic is often divided into two branches: inductive and deductive. In the philosophy of science, and the literature about research design, the terms inductive and deductive have sometimes been misused, and it is a confusing job trying to match up the descriptions and examples of the two that you might find in different sources. The following points might help.

Inductive reasoning

Inductive logic is concerned with the soundness of making inferences for which the evidence is not absolutely conclusive, and inductive reasoning is the process of reaching those inferences. If a body of evidence points in a particular direction, inductive logic could be applied to judge the strength of the inference or conclusion to which the evidence points. For example, if the accused in a murder trial has motive, opportunity, no alibi, his footprints at the scene and his fingerprints on the strangled victim's neck, inductive logic would lead us, through inference, to suspect the man's guilt. Our inductive logic suggests that the conclusion is probably correct, although the proof is not incontrovertible. Although there is no conclusive proof that this is the murderer, there is a weight of evidence in favour of that conclusion.

It is characteristic of inductive reasoning that from a series of individual observations (for example the fingerprints, the footprints, the
dodgy alibi) each of which is 'true' but which are not individually
conclusive, we put together a body of evidence from which we arrive
at a general conclusion.

Many descriptions of science have suggested that inductive reasoning is the 'scientific method'. Certainly there are many examples of science seeming to proceed by
induction. If we make an observation over and over again, and it always turns out the
same way, then we reach a general conclusion on the basis of a number of our individual observations. For example, if we see the sun come up in the east every day year
after year, we form a belief that the sun always rises in the east. When Galileo dropped
a number of different objects from the top of his tower and found that they all fell at
the same rate, he inferred a general truth from his specific observations, and it became
a 'law of nature' that all bodies fall with a particular acceleration under the influence
of gravity. For all Galileo could say, there might one day be an object that falls at a
different rate; likewise, the sun might come up somewhere else eventually, or new
evidence might acquit the alleged strangler. But the weight of observations so far allows
us to infer our conclusion. Suppose someone asks you if a dissertation needs to have
a good reference list to get a first class mark. If you say that all the first class dissertations you've ever seen have clear reference lists at the end while some of the third class
dissertations have not, and that you therefore believe that you do need a reference list
to get a first class mark, then you are employing inductive reasoning. You cannot be
convinced that your conclusion is undoubtedly correct, but the evidence supports it.

To illustrate an inductive approach to research you might consider Mercer's (1999)
study of the role of NGOs in Tanzania. In this study, Mercer sought to answer the
question 'Are NGOs making a difference?'. She presented a range of evidence (that
NGOs were disproportionately represented in Dar es Salaam and the richer parts of
the country, that they were not independent of the state, that attempts were being
made to co-opt the local NGO sector into social service provision, etc.), such that
she was able to conclude that 'the NGO policy is unlikely to facilitate structural
and sustainable changes in state–society relations' (p. 252). Many studies in geography
and in other disciplines employ an inductive approach.

Deductive reasoning

Deductive logic is concerned with the rules for determining when an argument is valid,
and it is thus concerned with establishing conclusive inferences, or what we might call
in common speech 'definite' conclusions. In other words, from our point of view, it's
about whether or not your argument can show if something really is true or false.

Deductive logic is not interested in the weight of evidence that leads to an
inference, but with the validity or conclusiveness of the argument.

Let's go back to the example of reference lists at the end of the previous section. If you argue that dissertations must adhere to the departmental guidelines in order to get a first, that the need for a reference list is stressed in the guidelines, and that you therefore conclude (deduce) that you need a reference list to get a first, then you are employing deductive reasoning. If your premises are correct, then your reasoning soundly and inevitably leads to your conclusion. Whereas inductive reasoning can be said to work from the particular to the general (for example from the particular instances of objects falling to the general law about falling objects), deductive reasoning can be said to work from the general to the particular. With the example of the reference lists, inductive reasoning begins with the particular instances of dissertations you have read and works from those instances to an inference about dissertations in general. Deductive reasoning begins with the general rules about dissertations and works from that general framework to the specific issue of your particular dissertation.

Deductive logic seeks to differentiate between valid reasoning and invalid or unsound reasoning. In common usage when someone says 'that's illogical' about something, they usually mean it is deductively unsound. One of the best-known forms of deductive reasoning is syllogism. A syllogism is a formal argument consisting of two premises and a conclusion. For example:

Example 1
Proposition 1 (premise):   All ice-cream sellers wear white hats.
Proposition 2 (premise):   Tony is an ice-cream seller.
Proposition 3 (conclusion): Tony wears a white hat.

In Example 1, if the two premises are true, then the conclusion cannot be false, and the argument is therefore valid. Of course the premises may be false (for example it might not be true that all ice-cream sellers wear white hats), in which case the conclusion would also be false, although the logic is sound. We could think of the whole argument as being 'If all ice-cream sellers wear white hats, and if Tony is an ice-cream seller, then Tony must wear a white hat'.

Example 2
Proposition 1 (premise):   All ice-cream sellers wear white hats.
Proposition 2 (premise):   Tony wears a white hat.
Proposition 3 (conclusion): Tony is an ice-cream seller.

In other words: 'if all ice-cream sellers wear white hats, then if Tony wears a white hat, he must be an ice-cream seller.' In Example 2, even if the premises are true the conclusion might be false, because other people as well as ice-cream sellers can wear white hats. The argument is clearly unsound. There are all sorts of mistakes that can make an argument unsound. Many, such as circular reasoning (e.g. 'I believe in God because the Bible says He exists and I believe in the Bible because it is the word of God') are familiar in everyday usage and easy to identify. When you are dealing with unfamiliar subjects in your research where the 'answer' is not intuitively obvious, you need to take great care to ensure the validity of your argument. Any good textbook on logic will provide you with a substantial list of logical fallacies that must be avoided in your work.

The following list shows some examples of faulty logic (even though the premises and, in some cases, the conclusions are true, the reasoning is invalid) and of sound logic (even though some of the premises here are false, the reasoning is valid if we assume the premises to be true):

Faulty

1   All men are human beings.
    All women are human beings.
    Therefore all women are men.

2   All cats are mammals.
    No dogs are cats.
    Therefore no dogs are mammals.

3   No dogs are birds.
    No birds bark.
    Therefore no dogs bark.

4   All head-hunters are primitive.
    Some Londoners are not headhunters.
    Therefore some Londoners are not primitive.

5   All people are mortal.
    All mortals are fallible.
    Therefore some fallible things are not people.

6   Socrates is a man.
    Man is a species.
    Therefore Socrates is a species.

Sound

1   All idiots are happy.
    All geographers are idiots.
    Therefore all geographers are happy.

2   Some geographers are stupid.
    No one who is stupid is wise.
    Therefore some geographers are not wise.

3   All geographers talk sense.
    No politicians talk sense.
    Therefore no geographers are politicians.

You need to be able to recognize good from bad reasoning in your own work and in situations where the right answer is not immediately obvious. How about:

1   Some pearls are not white.
    All white things are beautiful.
    Therefore some beautiful things are not pearls.

2   Some eskers are made of sand.
    This ridge is not made of sand.
    Therefore this ridge is not an esker.

Now put your own research arguments into this style and see if they are really valid.

In constructing an argument where a conclusion follows from premises, your argument must be logical and your premises must be sound. If your argument is logical then the strength of your conclusion depends on the strength of your premises. In your research you will need to accept certain pieces of information as being true without questioning them. This is part of the framework in which your research is set. For example, if your dissertation involves use of the gravitational constant, you will assume that its value is $9.81m\ s^{-2}$. These are your assumptions, and the strength of your conclusion is limited by the strength of those assumptions. However, your conclusions are also coloured by your theoretical or philosophical standpoint, for example by your willingness to make value judgements and, if you do so, by your personal or professional values.

Consider the following:

1 Cats kill birds.
   People like birds.
   Cat-owning should be discouraged.
2 Cats kill birds.
   People like birds.
   People should be encouraged to dislike birds.
3 Cats kill birds.
   People like birds.
   Cats should be discouraged from killing birds.

The same observations can inspire different policy decisions in different observers. It is important to be aware of where the boundary falls between logic and judgement, and whether your branch of the discipline expects research to cross that boundary.

To illustrate a deductive approach to research you might consider Knight's (1989) study of the way that layers of clean ice and debris-rich ice are entrained into the base of a glacier. Knight set out specifically to find out whether the idea that both types of ice were entrained by the same freezing mechanism, as had been suggested in previous literature, was actually right. To do this, he identified a specific characteristic (the isotopic composition) that would be the same in both the clean ice and the debris-rich if, and only if, both ice types had been formed by the same freezing mechanism. Before he even collected the data, Knight could say that if, and only if, his measurements showed isotopic differences between the ice types, then the idea that they formed in the same way could be thrown out and replaced by the conclusion that they were formed in different ways. They did, and it was. Many studies in geography and other disciplines employ a deductive approach.

The logic of (scientific) discovery

At this point, you should ask yourself what branch of logic you will be using in your dissertation. Do you aim to reach your conclusion on the basis of a body of evidence in support of your argument, or are you aiming to reach it by testing deductions from your argument?

The philosopher Karl Popper (Popper 1959) drew together some of the points that we've been talking about in the last few sections into a framework for research which a lot of geographers, especially physical geographers, have found useful. It is referred to variously as 'critical rationalism', 'the hypothesis-testing approach' and 'the hypothetico-deductive method'.

The background to the approach is that inductive reasoning cannot provide conclusive arguments of the type that research demands. Popper even goes so far as to argue that there is really no such thing as inductive reasoning. According to the critical rationalists, scientific research relies on deductive reasoning, where conclusive arguments are possible.

One of the best-known examples used to defend and explain the critical rationalist position involves the question 'Are all swans white?'. One way of answering the question (one research design) would be to look at all the swans you could find and observe their colour. The more white swans you found the more convinced you would be that all swans were white. When you had seen enough swans, or when you ran out of time or funding, you would conclude that since you had seen lots of white swans and none of any other colour all swans were indeed white. We can recognize that approach to be what we called inductive reasoning earlier in the chapter. We have not conclusively shown that all swans are white because we have not seen all swans, but we have a substantial number of confirming instances, a body of evidence that supports the conclusion. Nevertheless, if we had to answer yes or no to the question 'Are all swans white?', we could not do it on the basis of our research. We could only say 'probably' or 'that might be true'. In effect we have made no progress because we knew it might be true before we started. On the other hand, if we were to find a black swan, we would know for sure that not all swans were white. A single observation allows us to reach a definite answer to the question. The point is that we could never get a definite 'yes' unless we looked at every single swan in the world. For most research problems we could never be sure that we had looked at every subject, so it is impossible ever to answer 'yes'. By contrast it is possible to reach a definite 'no' if we find just one piece of evidence that falsifies the suggestion made in the question.

The implication is that it is a waste of time designing a research programme to find evidence in support of an idea because no amount of supporting evidence can 'prove' a point.

The sensible thing is to design a research programme that tries to find evidence against your idea. If you don't find evidence against the idea, then you are in a position, like the inductivist's, of having to say 'maybe'; except that your 'maybe' could be stronger than theirs if you have made a positive effort to look for black swans while they just looked at any old swans and wondered if a black one would turn up. If you do find the black swan, which you are more likely to do if you have gone out to look for one than if you haven't, then you are in a very different position. You can say with confidence that the answer to the question is 'no, not all swans are white'.

The logic involved in the falsification is deductive. If all swans are white, goes the argument, then we should not be able to find any swans of any other colour. In fact we have found a swan of a different colour, therefore we conclude that the premise is wrong and not all swans are white. Rather like the syllogisms we discussed earlier we can recognize a formal structure to the argument. First, there is a hypothesis. This is a possible answer to the question; in our case the hypothesis would be that all swans are indeed white. Second, there is a prediction. If all swans are white as we hypothesize, then we will not find any non-white swans. This allows us to formulate a test of the hypothesis: go looking for non-white swans. Third, there is the observation: 'Look, a black swan in Australia!' Finally there is the conclusion, the result of the test is that the observation falsifies the hypothesis so the answer is 'no'. The alternative result of the test would be the observation that we could only find white swans and couldn't find any non-white swans. In that case our conclusion would be that indeed all swans might be white, but we couldn't be absolutely sure.

Popper's approach works well for some questions (Are all swans white? Definitely not), but rather less well for others (What colour are swans, then? Don't really know). Nevertheless, it contains some things of value whatever your research question because it encourages you to take a look at the answer to your research question in the manner of the very meanest examiner.

Designing your own research

By now, we hope, you are beginning to see that getting an answer to your question, or doing your research, involves more than just making some measurements and hoping they 'give you an answer'. If you design your research badly, someone (the examiner) will say 'you can't reach that conclusion from that evidence'. We've already explained that your research design depends on your question, and looked at some of the basics of research design, so now it might be helpful to look at some specific applications of different research designs to different questions. In a short book like this we don't want to cover every possible approach to every possible question. We simply want to demonstrate how important research design is to answering research questions.

In the next section we give a point-by-point guide to the approach Popper advocates. If you decide that this approach is suitable for your project, then you can use the guide to lead you through the stages of your research. If you decide that this approach is not suitable, then thinking about why it isn't so, should set you on the path to a more suitable research design. We think, however, you will probably still find that at least some of the basic framework applies to your work, or that a part of the point-by-point guide is relevant to your particular project. Certainly this approach is a sound one for many types of project, and if you can use it you will be on firm ground.

Eric says: Your examiner will spot a dodgy research design quicker than the fashion police will spot a fake designer T-shirt.

Research design for explanation: finding out 'how and why'

This guide is based on the hypothetico-deductive approach that we described earlier. Here we use a more 'geographical' example than the swans!

Step 1   What am I trying to find out? The question

Scientific investigation in general can be broken down into different stages. First, there is discovery: someone finds something new, be it a country, a physical feature, a process or a relationship. Second, there is description: saying exactly what the new discovery consists of. Third, there is explanation. For dissertation writers, explanation is generally the most interesting part of the investigation to take on board. It's the 'how?' or 'why?' part of finding out. There are many different specific questions that could fall under this heading, and several different types of question, but many of them can be reduced to the form 'Why is this like it is?'.

Step 2   What is it like? The preliminary description

In order to start explaining why something is as it is you need to know exactly what it *is* like; in other words you need to know exactly *what* you are trying to explain. For your own preparation, and in your report, you will need to describe your subject (the feature, process or whatever). The nature and detail of your description will depend on the nature and complexity of the subject. You might use maps, diagrams, photographs or statistical data as part of your description. The aim is to give yourself, and eventually your readers, a clear (detailed and unambiguous) idea of what it is you are trying to explain.

Step 3   How can I begin to explain? The hypothesis

The way to begin is to come up with some possible answers to your question, in other words some *hypotheses*. In many cases, your hypothesis is no more than a reordering of the words of your research question. To go back to the examples we gave in inductive and deductive logic, Mercer's hypothesis was 'NGOs aren't making a difference', and Knight's hypothesis was 'debris bands are entrained by the same freezing process as the clean layers of ice'. You might be able to come up with hypotheses from your own imagination, but it is important also to explore the existing literature on your topic. Find out what other researchers have suggested about your question. It might be that someone has published what they think is 'The Answer', or a number of researchers might have put forward several different hypotheses. It is very important that your dissertation considers hypotheses that exist in the literature, to set your work clearly into an existing scientific context, even if you also move on to consider new hypotheses that have not been proposed before. At this stage it is a good idea to come up with several alternative hypotheses.

Step 4   How can I test these hypotheses? The predictions

Predictions take the form: 'If that is true . . . then this must follow.' Every hypothesis, or possible answer to your question, has a set of implications or expectations associated with it. For example, if you wonder what a particular room in the geography department is used for, and you suspect (hypothesize) that it might be a lecture room, then you would expect (predict) that if you put your head around the door you would see rows of seats facing a stage or a lectern. If your question is about the origin of a scattering of boulders on a hillside, and your hypothesis is that the boulders are a rockfall deposit, then your prediction might be something like: 'If this is a rockfall, I would expect the largest boulders to be at the bottom of the slope and the smallest ones at the top.' In order to arrive at your prediction, you need to know something about the kind of feature you are interested in. In our examples, the predictions rely on us knowing some of the typical characteristics of lecture rooms and rockfall deposits so that we can compare our observed feature (Room 207 or the scree slopes at Wastwater, for example) with our theoretical model. For the subject of your dissertation, you will probably need to do a lot of background work in the library to enable you to come up with sensible predictions. Bear in mind, of course, that to be useful for your research your prediction has to be *testable*. It's no good predicting what the room will look like if you can't look inside to check, for example, and it's no good predicting boulder size distributions if you can't go out and measure the boulders.

Step 5   Will any testable prediction do? Risky predictions

Some predictions are more useful than others. For example, if we argue that 'If it is a lecture room there will be chairs in it', and if we find that there are indeed chairs in the room, we still won't be sure that it's a lecture room because other types of room *also* have chairs in them. What we need is a prediction that will be true *only* if the hypothesis is true. You should try to identify 'risky' predictions. These are predictions that are very unlikely to occur unless your hypothesis is correct. You might find that you can think of no single characteristic of your hypothesis that could furnish a risky prediction to test. It might be that you need to consider an *assemblage* of characteristics. For example, lots of rooms have chairs in them, but only lecture rooms have chairs, a lectern and a slide-projector. In some circumstances you might have to consider situations where there is no single characteristic that is always true of the feature you are interested in. Statistical analysis might help here (Chapter 7). If you construct your prediction with specific statistical tests in mind, you might be able to test your hypothesis on a probabilistic basis.

Step 6   What do I do with my risky, testable prediction? The test

Once you have a good prediction, go and test it. This will involve making some kind of observation. In our examples, it could mean looking into the room or measuring boulder sizes on the hillside. Your observations might take the form of fieldwork, laboratory experiments or documentary searches, but will certainly involve some sort of data collection. It is very important that you know *before you start your data collection* what observations or results are possible, and that you know exactly what each of those results would mean for your hypothesis. For example, your earlier library work might have indicated that all rockfalls have the biggest rocks at the bottom. You will then be able to say *before you do your fieldwork* that if the biggest rocks are not at the bottom of your hillside, then your feature is not a rock-fall. You will also know *before you start* that if the biggest rocks *are* at the bottom then your feature *might* be a rockfall. Therefore you should know, before you make any observations, exactly what you *need* to measure.

Step 7   What if I falsify the hypothesis? Try another!

If you falsify your hypothesis, this means that it was not the right answer; the room is not a lecture room or the slope deposit was not a rockfall. This is progress. In everyday life there is a tendency for us to be disappointed if our ideas turn out to be wrong. This is not the case in science. Finding out that an apparently reason-able answer is in fact incorrect is a major achievement. However, at this stage you must be *quite certain* that your observation falsifying your prediction really has falsi-fied the hypothesis. Check once again that the prediction *had to come true* for the hypothesis to be verified. Is there any chance that the hypothesis could be true even though the prediction was not? Could the cleaner have moved the chairs out of the room, or could the rocks have been disturbed by human activity? Of course, you should have thought of that when devising the test (Step 5, above). If you are confident that you have falsified the hypothesis, the next thing to do is to return to Step 3 (above) and develop a new hypothesis to work on.

Step 8   What if I fail to falsify the prediction? Try again!

If you fail to falsify a hypothesis, that means that you *might* have the right answer; the hypothesis *might* be true. On the other hand, as we explained when we were talking about deductive logic earlier, it might be that your hypothesis is in fact incorrect but that the test you used was unable to show it. Rockfalls are not the only process that produces deposits with the largest rocks at the bottom, so finding that your deposit has the largest rocks at the bottom doesn't prove that it *must* be a rockfall. Therefore the next thing to do is to think of *another* test of your hypoth-esis, to try again to falsify it. Think of another risky, testable prediction based on your hypothesis, and return to Step 4 (above). If you have trouble thinking of another good prediction, remember that you can use assemblages of related predic-tions; none of the predictions independently might be 'risky', but taken as a group they might be.

Step 9   Is there no way to progress beyond Stage 8?

When you falsified your hypothesis we sent you back to try a new one, and when you failed to falsify that hypothesis we sent you back to try again. It might seem as if this method of research leads you round and round in circles with no satis-factory conclusion. In fact there will come a point where you run out of sensible hypotheses to test, and where you run out of good tests to apply to the hypotheses that you have failed to falsify. At this point, all your hypotheses will have been either falsified or tested until you run out of tests. We have already said that falsification represents a sort of progress, but what about the hypotheses that you have been unable to falsify?

Step 10   If my hypothesis isn't false, is it the right answer?

Even if you have made a number of risky predictions and they all turn out to be true, you still cannot be sure that your hypothesis is correct. Remember the swans; however many white ones you see, you can't be sure that they're *all* white. This is not as depressing as it might seem. The whole of scientific 'knowledge' is made up of nothing more than hypotheses that have not yet been falsified. If you have tried very hard to disprove an idea, and not been able to do it, then you might well be happy to accept the idea for the time being. Perhaps one day someone will think of a new prediction, or new technology might permit a new test, and your 'answer' will then be rejected. The physics of Isaac Newton was accepted for nearly 300 years before Einstein falsified it! Much of Einstein's physics is still accepted, and seems to 'work', but may one day be falsified and superseded. That's the nature of scientific progress. If, at the end of your research, you retain an unfalsified hypothesis, then you might let it stand as your 'best' provisional answer, and put it forward as a tentative explanation.

Research design for other questions: finding out 'when, where, what . . .'

There are many different types of research problem that you might choose to tackle in your dissertation. We have already argued that most of these can be reduced to a very small number of types of question. In the same way, there are many different approaches to research design. For example, you may have learned about the differences between qualitative and quantitative research, or between historical and contemporary studies. Just as the range of research questions can be reduced to a small number of basic types, so the range of research approaches can be reduced. The essence of any research design is an approach to finding something out. Therefore it is essential, whatever your approach, to identify your question ('I am trying to find out . . .'). Whether you are adopting a qualitative or quantitative approach, you will need to collect data to answer your question, so you will need to know what data to collect ('In order to find that out, I need to know . . .'). You will probably find that, whatever the details of your study, the framework that we have given in the previous section ('Research design for explanation') will serve you well. You can accommodate differences in the style of your data or the context of your question quite easily while retaining the clear (dare we say 'fool-proof?') structure that will keep you on course. There are alternative approaches, but we have found that this basic structure is nearly always appropriate. By all means pick and choose the parts that are relevant. For example, if you are doing a purely descriptive study, then you need only use the first few stages of the method.

Remember, the purpose of your research design is to ensure that your work leads not just to an answer, but to an answer (to the question you asked) that is sound and valid. Be sure to follow a design that will do that.

Chapter summary and conclusion

1   There are lots of different ways of finding things out and doing research, but not all of them will work for your project, so you need to choose carefully.

2   You need to know exactly what you are trying to find out before you start. Understand exactly what kind of question you are asking, and be sure you know what the answer should look like. Are you looking for a description or explanation? Is your answer going to be a history, a reason . . .?

3   Follow a logical procedure for reaching the answer to your question. If you follow the hypothesis-testing approach that we recommend, be sure to follow the sequence of:

- questions
- hypotheses
- predictions
- observations.

Suggestions for further reading

Best, S. (2003) *A beginner's guide to social theory*, Sage Publications: London, 280 pp.

Hubbard, P., Kitchen, R., Bartley, B. and Fuller, D. (2002) *Thinking Geographically*, Continuum: London, 275 pp.

Kitchen, R. and Tate, N.J. (2000) *Conducting Research into Human Geography*, Prentice Hall: Harlow, 330 pp.

Rhoads, B. L. and Thorn, C.E. (eds) (1996) *The Scientific Nature of Geomorphology*, John Wiley & Sons: Chichester, 481pp.

What to do after reading Chapter 5

Professional researchers applying for money to carry out a piece of work are normally required to put forward a substantial research proposal that explains what they plan to do. Most institutions require dissertation students to produce something similar, but on a much smaller scale, as an early indication of what their dissertation is going to be about. It will be very useful for you to produce a detailed research proposal at this stage, before you start your data collection. Thinking through your plans in your head is OK, but most of us are very good at fooling ourselves into thinking that we are better prepared than we really are. The only way to be sure that you really know what you are going to do, and that you have avoided the major pitfalls, is to produce a coherent written plan. To ensure that professional researchers have done their planning and research design properly, grant-giving bodies usually have special research proposal forms to be filled in. You should now make sure that you can do the same.

Produce a research proposal of about 800–1,000 words (about two sides of typed A4) under the following headings:

1   Title of project *(What is the dissertation called?)*
2   Aim of project *(What are you trying to find out?)*
3   Scientific background and justification *(Why is this important and what work has been done on it before?)*
4   Methodology *(How are you going to do it?)*

5 Timetable of research *(When are you going to do what?)*
6 Location of data sources *(Are you going to use a Record Office, a river or an interviewee?)*

Different versions could include the headings: 'Data required', 'Data sources' and 'Data collection methods'.

If you really know what you are doing, you should be able to make a coherent and convincing case for your project on one side of A4 paper. Indeed, many institutions may require you to do this. Keep redrafting your research proposal until everyone you show it to can understand what you are doing and why you are doing it, and believes that your approach will work. Don't start your data collection until you've got this bit right. Check whether *your* institution has a form, and make sure that you can fill it in!

6

WHAT KIND OF DATA DO
I NEED AND HOW DO I
GET THEM?

This chapter explains the relationship between the problem you are
investigating and the type(s) of data needed to solve it; it looks at issues
associated with data collection; it distinguishes between primary and
secondary data; and it assesses the advantages and disadvantages of each.

Data, data analysis and data quantity

The type of data that you will need to undertake your research, what you will need
to do with your data when you have collected them and how many data you will
need to collect are three interrelated issues that you need to consider all together.
For the sake of this book, we have chosen to deal with these three issues separately
in this and the following two chapters. However, we should stress that you need
to address them all simultaneously and *not* sequentially. For example, if you are not
sure about the difference between qualitative, quantitative and spatial data feel free
to skip to the section in Chapter 7 where we deal with this.

What kind of data do I need?

Once you have reached this stage, and you have completed your research proposal,
you should know the data that are required in order to answer the question you
have set yourself and/or test the hypothesis(es) you have defined. You should know
whether the data you will need will be qualitative and/or quantitative.

**It is vital that before you start collecting any data you are absolutely
certain that these are the data you need.**

You will know whether you need to have measurements of windspeed (quantitative data), attitudes of residents in a town (qualitative data), evidence of an event in
the past that you think may have caused a present-day distribution of vegetation (could
be either qualitative or quantitative), or the distribution of Anglo-Saxon buckets
(spatial data). You will also know whether your dissertation is going to involve you
in fieldwork, laboratory experiments, questionnaires, computer modelling, collection

of data in a Record Office, or whatever. Take some time now to check that you do, in fact, know all this. Good research projects can come to grief here. It's all too easy to *think* that the data will give you the answer you need when, in fact, they won't. In Chapter 4 we asked you to list the questions you were going to ask and also to list all possible answers to those questions and what those answers would tell you. Now imagine that you have collected the data that you plan to collect. Will these data give you (or lead to) the answers to those questions? Are these all the data you will need? Do you need all of them?

If you are unsure of the answers to those questions, this and the following two chapters may help you to sort things out. You may also have realized that there may be more than one sort of data you could use to answer your question(s) and you may be wondering which sort to use. In this chapter we will look at the different types of data and the advantages and disadvantages of each.

Examiners' favourite viva question:

Why did you need to collect the samples of . . . when you already knew that . . .?

Types of data

Before we start talking about types of data, we may as well clear up an issue that crops up a lot with data. 'Data' is the plural of 'datum'. So, in correct usage, data should always be referred to in the plural (as we do here). You will find many instances where data are referred to as though they were singular, e.g. 'the data shows', instead of 'the data show'. Some people won't be too fussy whether you get this right or not. Your examiners, on the whole, will be.

Data are often divided into *primary* and *secondary*. Like most things, this division is not as straightforward as it might appear. For our purposes we will define primary data as those that you collect yourself. So, for example, measurements that you make of stream velocity in the field, questionnaires that you administer to visitors to a beach and measurements of infiltration into soil samples in a laboratory are all primary data. Secondary data, on the other hand, are data that somebody else has collected and that you will use for your own purposes. So, the age structure of the population of a parish that you obtain from somebody else's published work or stream discharges that you obtain from the National Rivers Authority are secondary data. The distinction between the two is that for primary data you will have first-hand information that will help you assess the reliability of the data, whereas for secondary data you won't. You will have no idea of the conditions under which the data were collected or of the diligence of the collector.

However, the distinction we have made here is flexible, even within geography. In historical geography, for example, public records, parish registers, etc. are regarded as primary data and the term secondary data is used to refer to data that you may obtain from other studies that are contemporary with the data. For example, census data is primary data for a historical geographer while data obtained from a study of agriculture in the 1850s, although originally based upon a survey at the same time, would be regarded as secondary.

One thing you should have checked before you reached this stage of your project is the data requirements of your institution. Some institutions, particularly in the past, have insisted on students undertaking some primary data collection, so that even if many of the data that you need can be acquired from secondary sources, it may be necessary for you to collect some original data yourself. For example, you could administer a questionnaire to supplement census data, or carry out a river survey to supplement hydrological data provided by the Rivers Authority. With more secondary data becoming available from such sources as remote sensing satellites and more extensive censuses, and an increasing emphasis on the role of a dissertation in evaluating analytical skills, the requirement for primary data collection is becoming less common. Nonetheless, it is a good idea to make sure. You don't want to have all your hard work downgraded and miss out on a prize because it falls foul of some archaic rule that even your tutor had forgotten about.

If you are not sure about the rules of your own institution regarding primary and secondary data, find out now. Read the handbook and check the web page.

Many dissertations will use both primary and secondary data even if, in some field-based projects, the secondary data amount to no more than using a map to locate the sites. Whatever type of data you use, you will need to give some consideration to their reliability.

Data reliability

An assessment of the reliability of your data is important because it affects the validity of the conclusions you may reach. In many areas of geography little regard appears to be given to the issue of data reliability. It is just assumed that the data are reliable. For example, it is not common to find discussion of this topic in research reports in physical geography, unless there are special circumstances that may have significance for the reliability of the data. On the other hand, historical geographers, for example, are very concerned about the issue and it will usually merit discussion in papers on historical geography. However, even if there is little public discussion of the topic in your particular branch of geography, that doesn't mean you can ignore it in your dissertation. In whatever branch of the discipline you are working, it is a good idea to make an objective assessment of your data before you draw conclusions.

In historical geography, for example, two issues are important. The first concerns the veracity of the data. Is the document you are using a forgery? The second issue arises from the fact that in historical geography greater use is made of surrogate measures than in other parts of the discipline. For example, if you wanted to study the effect of land ownership on the use of agricultural fertilizers in the nineteenth century, you might decide to use records of land tenure and sales of agricultural lime. This would not be a perfect measure of what you wanted to investigate so you would need to discuss its shortcomings.

Primary data

Advantages

The main advantage (see Box 6.1) of collecting primary data is that you have an intimate knowledge of them. That includes knowing how unreliable some of them are. All data are subject to measurement error of one sort or another. On the whole, however, we don't say much about the errors and there is a danger you may think that everybody else's data are much better than yours. They may be! But there's nothing like trying to make the measurements that somebody else has reported making to discover just how error-prone they are. Such an attempt can give you new insight into the conclusions that have been reached on the basis of the measurements! In Chapter 4 we used the example of a dissertation that examined grike depths and showed that one of the reasons behind this study was that the author had concluded that the existing literature, which was based on measurements of width, may be unreliable because of his own experience of the difficulty of obtaining reliable width measurements.

Box 6.1 Advantages and disadvantages of primary and secondary data

Primary data	Secondary data
Advantages	*Advantages*
Specific to the problem	Potentially large data set
Intimate understanding of their reliability	Often in computer-compatible form
	Quick to obtain
Disadvantages	*Disadvantages*
Small data set	May not be exactly what you want
Tedious data collection	No information on data quality
Time consuming	May have restricted access/use

Knowing the relative (un)reliability of various sets of data that you have collected can be very useful when you come to analyse them. If things don't seem to make sense it may be because one set of data just isn't good enough for the purpose you need it for.

There is danger in thinking that your data are wholly reliable.

There is a danger that once your field data get into your notebook, or worse still, into a computer file, you will think of them as wholly reliable. Thinking back to the conditions under which they were collected can help a lot. Imagine the case in which your questionnaire survey seems to indicate that most of the population in your study area is retired, yet the census data claim that the number of retired people in your area is below the national average. Realizing, from the note you made of the day in your field notebook, that you undertook most of your questionnaire work in the High Street on a Thursday (the day many retired people collect their pensions) might indicate that you cannot relate your census and questionnaire data. Such a realization might be a disaster for the project you had designed, but all may not be lost (see Chapter 10).

A second major advantage of primary data is that (within the constraints of what is technically and logistically possible) you can get exactly what you need for your project. One of the problems of using secondary data is that you can identify a question, but often the data have not been collected in such a way that they can be used to answer that specific question. They may have been collected too infrequently, at the wrong time or for the wrong administrative units.

Disadvantages

The main disadvantage of primary data is the time taken to collect them. Data collection is, on the whole, extremely slow. Furthermore, there is usually a marked learning curve so that initially data collection will be both slow and inaccurate. You are quite likely to discover that once you start, things aren't as you imagined they would be. Your planned data collection programme may not work. This may not become apparent for some time so that you find yourself throwing away a day's data because you come to realize by the end of the day that the methods you are employing are not sufficiently consistent. Added to this, your inexperience may well lead you to make mistakes. For example, you may be undertaking some surveying with an unfamiliar theodolite and fail to realize that, whereas the one you used before automatically levelled the vertical circle, this one requires you to do it manually. Everybody who collects primary data, even the most experienced, encounters these problems. So don't feel that you are unusually stupid. The trick is to know it will happen and allow time accordingly. If you are exceptionally lucky and everything does go according to plan, you may have some time to spare. Take a holiday.

One way to try to overcome this disadvantage is to undertake a pilot study. Again, of course, this takes time. But it is likely to be time well spent. So, whatever your data collection involves, decide first on your method of collection. Then go out on the first day and collect data in this way expecting to find things will go wrong. If you can, it is also a good idea to do some preliminary analysis of the data. Even if this is not possible, go through the steps you know you will need to perform and check that they will work. If your project involves using your field data to calculate something, have you measured everything that is needed? It will soon become apparent where the faults are. You may find that it's simply a matter of not having enough hands to hold all the equipment you need, and that you can solve the problem by hanging some of it around your neck. But you may find that your plan leaves a vital piece of information missing.

A pilot study can help identify any problems with your proposed method of data collection.

A second disadvantage of primary data is that data collection can be tedious. Often, you will find yourself doing the same thing over and over again. While it's not quite true to say that when you've measured one infiltration curve you've measured them all, the benefits to you and your project from subsequent measurements very soon start to seem disproportionately small compared to the time taken. There is nothing that can be done to mitigate this problem. All that can be said is that if the research problem exists and the data to solve it don't, there is no choice but to collect them yourself.

Finally, and largely as a result of the two previous disadvantages, using primary data means you will end up with a very small data set, and almost certainly smaller than you had planned to have. It is important that you know the relationship between the amount of data you actually have and the amount you will need (see Chapter 8). Otherwise, you may find that, although the data you have are just what you need, you have insufficient of them to enable you to answer your question(s).

Fieldwork

Much primary collection involves fieldwork, whether this is undertaken in a city, in a shanty town, under a forest canopy or on a glacier. There are issues of safety, ethics and data protection that can be associated with data collection. There are guidelines on safety, produced by a number of agencies, governing fieldwork by students. In addition, your own institution will have its own document regarding safety in fieldwork. Make sure you have read at least one of these documents and that you take the necessary precautions. You are likely to be required to complete a risk assessment form and have this form checked and approved by your tutor. There are ethical issues involved, particularly when people are the subject of your investigation. Your university is likely to have an ethics committee that needs to approve research involving people or animals. It is unlikely that the research for your dissertation will need to be approved by such a committee, but your department should certainly have an ethics form for you to sign and may refer your research to a university committee if your tutor thinks it necessary. (This is only likely to be the case if your research is unusual and outside the experience of the staff of your department.) Finally, there is the matter of data protection. If you collect data about people in such a way that they can be identified in your data set, then they have a right of access to those data. Again, your tutor should know about the rules governing data protection and should be able to advise you.

By definition, fieldwork means you are away from your base and away from advice, technical back-up, spares, etc. Fieldwork does, therefore, require some careful planning both for what you intend to do and what you will do if things go wrong. Make lists of the equipment you will need in order to carry out your fieldwork. Identify those things that are essential and, for them, work out what you will do if they break, or you run out. This doesn't mean that you need to take along two spare theodolites! It does mean that you should have a plan of what you will do if it breaks – even if this plan is no more than to come back! You might be able to identify all the separate tasks you have planned for the period of fieldwork so that if something does go wrong with the equipment needed for one of them, you can work on something else in the meantime. Think about the weather. Even if you are quite happy to work in the rain, the people you want to interview on street corners might not be. Likewise, large changes in river discharges following rainfall might be unacceptable for your research design. A list of things to think about before setting out on fieldwork is given in Box 6.2.

Box 6.2   Things to think about before setting out on fieldwork

- List of equipment
- Spare/back-up equipment
- Does your equipment work? Do you know how it works?
- Plan of action (including timetable, logistics and transport arrangements)
- Alternative plan of action for when things go wrong
- Do you have any necessary permissions?

If fieldwork is carried out a considerable distance from your base then you may have to assume that there will be no opportunity to return. Any data you fail or forget to collect will not be collected and your project will have to manage without them.

Foreign fieldwork

Everything that we have said about fieldwork generally applies to foreign fieldwork, but more so. There are disadvantages. Not only are you isolated from your base but you may be in a remote setting where you have no idea what the Spanish for 'water-permanent marker pen' is, even if you could find somebody to sell you one.

Having a dissertation about an exotic location is one way of arousing your examiner's interest.

If you do decide to undertake foreign fieldwork you will need to make more careful arrangements than if your fieldwork is more local and/or you will need to be prepared to spend more time sorting things out when they go wrong. But if all this sounds off-putting, there are also advantages. Remember (from Chapter 2) that in writing your dissertation you need to make the examiners want to read it. Having a dissertation about an exotic location is one way to provide an initial high level of interest. There are also more tangible advantages to you. You may be able to tackle a problem that lies close to your own field of interest. A large number of students study courses on Third World topics. Foreign fieldwork may be the only way you can undertake a dissertation on such a topic. Outside of these academic benefits are social/personal ones. You may be able to have a holiday in your field area as well (as well – not at the same time!). You could try to organize a group of your friends/fellow students to travel to a joint distant location. Imaginative projects or a group of related projects that are to be undertaken in a foreign location could well attract funding from some source.

Funding for fieldwork

Some institutions provide funds for students to undertake fieldwork as part of their dissertations; others do not. If funding is available from your institution, you will almost certainly be aware of it, although there may be such things as travelling scholarships that are less obvious. Other sources of funding exist however. First, there is your tutor. Your tutor may be involved in a funded research project to which your dissertation might be able to contribute. Second, there are charitable organizations that fund research of particular types. For example, in the UK the Friends of the Lake District give small grants to students whose dissertations may contribute to our knowledge and beneficial use of the Lake District National Park. Third, there may be commercial organizations interested in the results of particular types of market research. If you really want to work on a project that is going to be expensive, it may be worth spending some time looking for funding. However, you have a limited amount of time to spend on your dissertation, so delaying a decision on what you will do while you wait for a decision about funding cannot be allowed to take up too much of your time. Don't use it as an excuse for not getting started.

Permissions

Wherever you undertake fieldwork, you are likely to need access to somebody else's property. This may involve obtaining access to a woodland or a farm or that you will have to knock on people's front doors. Wherever you need permission you should obtain it. Most people are quite happy to give permission in response to a polite request. They may be less accommodating if the request comes after the owner has discovered you on his land. If you are thrown out halfway through a week's data collection you will have wasted a lot of time. For fieldwork in some areas (especially national parks in many countries) a permit may be required from the local or national authorities.

Laboratory work

Not all primary data need be collected in the field. For some types of project the data collection can be done largely or wholly in a laboratory. Again, two words of warning: first, if you plan to base your dissertation on laboratory work, check that your institution allows this. Second, just as with fieldwork, there are safety issues associated with working in a laboratory. Check with your institution. There may be regulations affecting the times and conditions under which you can have access to a laboratory that will restrict the type of laboratory work you can undertake.

In some instances, laboratory work will be associated with fieldwork as part of your data collection. For example, you may have collected a peat core from a field site and then need to spend time analysing the core, or you may have collected soil samples and need to undertake physical or chemical analyses of these samples. In both these cases, the time you will need to spend in the laboratory will be considerably greater than that in the field. Alternatively, you could undertake your entire project in a laboratory. Increasingly, in many areas of physical geography in particular, it is becoming common to see published results of laboratory experiments (see, for example, Dunkerley 2003; Harris *et al.* 2003).

Computer laboratories

You don't have to walk around in a white coat if you want to work in a laboratory. Just as laboratory measurements and experiments are as valid as fieldwork for solving problems, so is computer modelling or GIS. If you have a particular interest in computers and/or have good programming skills, you might consider a dissertation that is strongly based on numerical modelling. We will say more about numerical modelling as an approach to problem-solving in Chapter 9. Even if you decide to base your dissertation around numerical modelling, you can still include some fieldwork. Field data can provide an input to GIS or modelling, and you may like to undertake a dissertation that combines the two.

Secondary data

Advantages and disadvantages

To a large extent, the advantages and disadvantages of secondary data are the converse of those for primary data: see Box 6.1 (p. 63). Secondary data have the major advantage of being more readily accessible, and typically can provide you with large data sets relatively quickly. In some instances, secondary data will still involve you in some tedious data collection or transcribing of data from the original source to your notebook. So you may spend almost as long in a library or record office as you would in the field or laboratory collecting primary data. Increasingly, however, many data sets exist in digital form so that they can be obtained for direct input to an analytical computer program or a statistical package. However, life may not be as simple as that. You may find that, although the data do exist in digital form, there is a hefty price to be paid for data in that form. You may also find that, the benefits of modern technology notwithstanding, you are still stuck in an office copying out numbers with a pencil.

Such close contact with the data isn't a bad thing. As with primary data, some 'feel' for data quality and reliability is useful and may be obtainable from original records, whereas it is not from a digital record. As we have already said, it is particularly important in historical geography to be able to assess the reliability of your data. The original records may contain useful clues. Other clues may come from reading the work of others who have used the data and who may have had better access than you may have to the original sources. In addition, there may be reports that go with the data which will tell you something about the accuracy and reliability. For example, if you use data collected by a remote sensing satellite, the data will have undergone some pre-processing before you receive them. Such pre-processing aims to make the data more readily usable by most of the user community. If the sensor had a tendency to miss out some data, these data may have been 'interpolated' from neighbouring data. Without access to information that tells you this is so, you will not be able to tell the difference between a 'real' value and an 'interpolated' one. Likewise, censuses contain interpolations where full data sets were not available. These interpolations are made to make the census more useful to the people for whom the census data have been collected. You are not among the most significant of the users of census data, so your needs won't have been paramount when the interpolation methods were devised!

So far, we have discussed secondary data that have been collected with a view to their use by a third party, for example census data. Other secondary data might not be designed for public access. Such data may have been collected by an individual for a particular research project and now you want to use the data for some other purpose. Treat such data with particular care. Where data have been collected with a user community in mind, some serious attempt will have been made to make sure they are as consistent and reliable as possible. Likewise, there will probably be some documentation on the data. For data that you might extract from the appendix in somebody's thesis or journal article, no such safeguards are likely to exist. Bearing in mind that the original researcher did not have your project in mind when the data were collected, and may, anyway, have been reluctant to draw attention to weaknesses in the data, you may well find that this apparent short cut to your data proves to be a time-consuming sidetrack.

Permissions (ownership of sites/copyrights)

There may not be free public access to secondary data, and even if there is, there may be restrictions regarding what can be done with them. In particular, publication of secondary data may not be allowed without permission. You will need to check on any restrictions of use that may apply.

Foreign data

Access to foreign secondary data can often be difficult. Basically, the problem is likely to be that you will not have the same information regarding the format and structure of the data source as you would have for a source in your own country. There is no reason to believe that data comparable to those which exist in your home country will exist elsewhere. It may be better elsewhere; it may be worse. Consequently, if you write for data your letter is unlikely to arrive on the right person's desk and even if it does, it may well ask for things that are not available in the form that you have asked for. Given that the person will not know the requirements of your project, he/she cannot be expected to be able to relate the data that are available to your needs. Unless you have a reliable route to secondary data in a foreign country, either from personal contact or through your tutor, it is very risky to trust the success of your dissertation to a foreign secondary data source.

Chapter summary and conclusion

This chapter has described the differences between primary and secondary data and looked at the advantages and disadvantages of each from the point of view of undertaking a dissertation.

What to do after reading Chapter 6

Read Chapters 7, 8 and 9.

7

WHAT CAN I DO WITH MY
DATA WHEN I'VE GOT THEM?

This chapter looks at the processing you may need to carry out on
your data to answer your question(s). It considers types of statistical
tests that are appropriate for different types of quantitative data and
processing methods for qualitative and spatial data.

Data processing

Your data will enable you to answer the question or test the hypothesis that you
defined in Chapter 4. However, the data alone are unlikely to be able to do that.
Collecting the data rarely leads you directly to the answer that you need.

For example, if your dissertation sought to test the hypothesis that new manufacturing industry was preferentially located in small, rather than large, towns and
your study showed that all the small towns you investigated had higher rates of
growth than any of the large towns you studied, then the answer to the question
would be immediately apparent and no further data processing would be necessary
in order to answer that particular question. But few data sets will be so unequivocal and, even if they are, you might wish to explore the question further. So, in
this example, having divided your towns into small and large and obtained the
answer that there is more new manufacturing industry in the small towns, you might
want to see if the relationship between new manufacturing and town size is stronger
than simply big–small and less–more. Is there a linear relation between growth of
manufacturing industry and town size, or is there a threshold of size below which
growth is rapid and above which it is slow? Alternatively, the results might have
shown some big towns with a lot of growth but it seemed to you that there was
a preponderance of small towns at the top of the growth league. How would you
test the reliability of this personal impression?

One way or another, therefore, you are likely to need some processing of your
data. Regardless of the methods you use and the type of data you have, all data
processing has the same purpose: it moves you forward towards providing the answer
to your research question. For convenience, however, we will divide data processing
into three types: quantitative data processing, qualitative data processing and spatial
data processing.

Quantitative data processing

There are statistical methods to suit every level of numeracy.

This may sound horribly like statistics, and it is! However, there are statistical methods to suit every level of numeracy. What's more, the level of statistical analysis you do depends very much upon you and your choice of subject area for your dissertation. Some branches of the discipline (cultural geography, for example) have a relatively limited tradition of statistical analysis. In others (for example, plant ecology) there is a strong tradition of employing statistical procedures in solving problems. If you don't like statistical analysis, then you may want to keep to the simple descriptive statistics and tests, and choose your subject area accordingly (or not have any quantitative data at all). But if you are good at, and like, this aspect of the subject, then you may want this part to be a major focus of your dissertation.

Types of quantitative data

For statistical analysis it is necessary first to know into which category your data fall. Four categories of data are recognized. These are nominal, ordinal, interval and ratio. Definitions of these categories, and examples of data from them, are given in Table 7.1. These categories are generally regarded as forming a hierarchical sequence with nominal data at the lowest level and ratio data at the highest level. The lowest level of information you can use in any form of statistical testing is some descriptive label that you can apply to your data, such as its colour. Generally, you can do more with your data from the point of view of statistical tests, the higher up the hierarchy your observations come. For example, you might be studying soil moisture levels and may have recorded the data in the field in the categories 'dry, damp, moist, wet and waterlogged'. Such data would not be equivalent to, or be amenable to the same statistical treatment as those you had collected and analysed in the laboratory to determine the percentage of water in the weight of the soil sample. One of the things you have to decide on is the level of data you need for your problem. As you go up the hierarchy, it usually takes longer to collect the data. So there is no point in collecting ratio data if all you really need is ordinal data. For example, assume you had decided to investigate the growth of vegetation on abandoned spoil heaps. The records of the construction of the spoil heaps provide you with a relative chronology (oldest, second oldest, etc.), but not their absolute ages. So for the ages of the spoil heaps you have only ordinal data. If you want to relate age to vegetation growth, then all you need is a similar level of information on vegetation cover (none, some, quite a lot, etc.). You can get data of this type very much more quickly than you could obtain percentage cover estimates using a quadrat.

All statistical tests have a minimum level of measurement for the data that can be used in them. For example, two tests that measure differences between two sets of data are the Chi-squared (χ^2) test and the Mann-Whitney U test. The χ^2 can

Table 7.1 Scales of measurement for data.

Data type	Definition	Example
Ratio	Data measured on a ratio scale have the property that they are referenced to a zero value so that two values retain the same ratio irrespective of the units in which they have been measured	Length, volume
Interval	Interval data are data where individual observations can be compared one with another without the need to refer to membership of classes. You can subtract one value on an interval scale from another and obtain a sensible answer. However, interval data are measured on an arbitrary scale and so are not absolute quantities	Temperature in degrees centigrade
Ordinal	Ordinal data are data that are arranged in classes, which themselves form an ordered sequence from lower to higher	Degree class
Nominal	Nominal data are data that are categorized by their name only	Colour, Gender, Species, Lithology

be used on nominal data, but the Mann–Whitney *U* test requires that the data are measured at least on an ordinal scale. You can always use a statistical test that works on a lower category of data on data from a higher level. Consequently, if you have ratio data, then all forms of statistical treatment are potentially available to you. However, if you do use on ratio data a test that would work on ordinal data, the test will not use all the information that is available in the data. For example, if you measure two sets of temperature data, then you can see if they are related using rank correlation, but this technique will use only the rank ordering of the temperatures and, therefore, not be as powerful as the product–moment correlation which uses the actual values that have been recorded. You should always try to use tests that require, and therefore make the most of, the level of measurement that you have. Conversely, you should collect data only at the level needed for the statistical analysis you plan to undertake. That's why you should always think ahead to the data-analysis stage of a project while you are planning your data collection.

Eric says: **Statistics might look scary, but it really is important to make sure that you understand the basic rules that determine which tests can be used for what.**

If your data fall into the category(ies) interval or ratio, you also need to know whether they are 'parametric' or 'non-parametric'. These terms are used to distinguish between cases where the distribution of values that you have conforms to some model distribution (parametric data) and where it does not. The commonest such model distribution is the normal distribution. Many statistical tests are based upon the assumption that the data are distributed in this manner. Consequently, the results of the test are valid only if this assumption is not violated. A very common error in weak dissertations is to use a test on data to which the test is not appropriate. However, having said that, many such tests are quite resilient to violations of the assumption. What's more, you will find countless examples in the literature where such tests have been used, even though the data do violate the assumption!

Examiner's favourite viva question: Why did you attempt regression analysis on data that were measured on an interval scale?

The important points are (1) not to be intimidated by the complexities of statistics and (2) to be aware of the rules that govern the correct use of a particular statistical test. Very often there are alternative statistical tests to examine a particular type of problem. For example, you can measure the correlation between two things using the contingency coefficient if the data are measured at the nominal level, rank correlation coefficients if the data are measured at the ordinal level, and the product–moment coefficient if the data are normally distributed and measured at interval or ratio level.

You should try to make sure that you use the test that suits your data. However, there may not be a statistical test available for exactly what you want to do. After all, statistical tests exist because somebody has already identified a statistical problem of a particular type and worked out a way of solving it. Some types of problem have not been solved because nobody has yet come up with a way of doing so. In such cases, you will have to use the best statistical technique there is, but be aware of the implications of the fact that it doesn't do exactly what you want. Don't worry too much about being quite blatantly disrespectful of the rules governing statistical tests. You may offend the statisticians but, on the other hand, you may come up with some interesting geographical ideas. Statistical tests are no more than a useful way of moving forward at a particular point in research. They do not define the end-point. If you feel unsure about the concepts discussed in this section, look at Box 7.1 in which we have listed some types of data. Can you assign these data to the categories nominal, ordinal, interval and ratio?

Samples and populations

Are you interested only in the data you have collected, or do you wish to regard your data as a sample drawn from a larger (perhaps infinite) population? Are you using your 200 interviews just to find out about the 200 people you interviewed or to reach a conclusion about people generally? The way you defined your question in Chapter 4 will have determined which of these is the case. It is a good idea now to make certain you are sure of this. In many forms of data processing, exactly what you do will depend on being clear about this point. Don't form the impression, easily reached from reading books on statistics, that it's somehow better or more normal if your study is about a sample of a larger population. Much of statistical theory was developed for such situations, and so many books on statistics tend to assume this to be the case. But that has nothing to do with the validity of your question, which depends upon quite

---

Box 7.1 What are the scales of measurement of these examples? The correct answers are given at the end of the chapter

Extracts from some data that may be collected for dissertations	Nominal, ordinal, interval or ratio?
20.3 m/s. 11.4 m/s, 7.8 m/s, etc. (measurements of windspeed)	
Detached, terraced, detached, apartment, detached, etc. (data on house types in a residential survey)	
6.9 ha, 18.4 ha, 37.2 ha, 14.3 ha, 4.7 ha, etc. (areas of fields)	
Fine, medium-fine, coarse, coarse, medium, etc. (textures of soil)	
67 per cent, 91 per cent, 34 per cent, 18 per cent, etc. (households with two cars in different enumeration districts)	
Arable, pasture, woodland, pasture, etc. (land use)	

---

different things. We will have more to say in Chapter 8 regarding sample sizes for studies in which the data are taken to be representative of a larger population.

Types of question and types of quantitative data processing

In Chapter 4 we showed how, although there is an infinite number of questions you could ask, these questions can be grouped into a small number of types. We gave generalized forms of these questions and then reduced these generalized forms to a succinct notation. In this chapter we will use these same succinct question forms to discuss data processing because the form of processing you can undertake is a function of the type of question. A modified form of Box 4.4 is repeated here as a reminder (Box 7.2). First, however, we need to say a bit more about question types in relation to data processing.

Box 7.2 represents something of a simplification in that it deals only with either univariate or bivariate questions. Thus the questions are about A or about A and B – what is A like; is A like B, and so on. The questions you have asked may not be this simple. You may have asked questions such as: 'Do the size, average gradient and vegetation of a catchment affect the shape of stream hydrographs?' 'Do the age, gender and social status of a person affect their economic status?' These questions are a bit like A $\rightarrow$ B?, but not quite the same. You could represent them as 'Do A_1, A_2, A_3, affect B?', or in its succinct form ' A_1, A_2, $A_3 \rightarrow$ B?' In our examination of the relationship between types of question and types of data processing we will assume that the succinct forms of question that refer to A, or A and B, may also be regarded as succinct ways of referring to A_1, A_2, . . . A_n and B_1, B_2, . . . B_n. Consequently, a question of the form A > B? is regarded as the same as A_1, A_2 > B?

Box 7.2 Types of question

Questions	Generalized form of question	Succinct form of question
What is the catchment area of the Potteries Shopping Centre?	What's A like?	A = ?
Do these two deposits of glacial till have the same provenance?	Is A like B?	A = B?
Do the schools in this town differ in terms of the socio-economic backgrounds of their pupils?	Is A different from B?	A ≠ B?
Does the record of annual expenditure on agricultural lime provide a better measure of rural wealth than the record of grain sales in eighteenth-century Shropshire?	Is A better than B?	A > B?
Does the distribution of oak in Britain correlate with the distribution of holly?	Are A and B related?	A ↔ B?
Does mean river channel width increase as discharge variability increases?	Does A affect B?	A → B?
Does moving to a large town increase your chances of dying from a respiratory disease?	Does A cause B?	A => B?

It will not be possible in the space available in this chapter to list all possible statistical techniques that are suitable for particular types of question. In particular, more sophisticated techniques will not be mentioned. However, the aim of the chapter is not to provide a potted statistics course. We will say nothing about the details of any statistical technique, because you can find those in any book on statistical methods. Rather, what we are trying to do is to show the relationship between statistical tests and types of question. When you look at a book on statistical methods you will find an array of techniques presented. The task you will most likely face is to determine which of these techniques will help you answer your particular type of question. The most important thing you need to know about any statistical technique is what exactly it does. At the end of this chapter you should be able to identify what statistical treatment you want to perform on your data. If that treatment has not been mentioned here then you should, at least, be able to describe what you are looking for. That way, it will be easier to find the right statistical treatment for your purpose, either by asking somebody or by looking in appropriate books.

Nowadays, there is no need to perform the tedious task of calculating values for statistics. Simple statistics can be performed using a hand calculator, and there are numerous packages available for computers that will allow you to undertake even quite sophisticated statistical data processing. It is, therefore, of fundamental importance that you don't lose control of your data processing to the statistics package and end up performing statistical analyses without knowing either why you are doing them or what interpretations you will be able to put on the results. You may think pages of statistical analyses will impress your examiners, but they won't unless they are matched by an equal amount of intellectual achievement. In fact, your examiners are more likely to downgrade your dissertation if you've performed lots of statistical analyses and then make no interpretation of the results than if you'd done a more modest amount of analysis.

A = ?

If your question is of this form, then it means your dissertation is about description. You are trying to find out what something is like. Consequently, your data processing will be an attempt to reduce the data you have collected to some sort of summary form that will enable you to provide a succinct description. Thus, if your dissertation is about visitors to a National Park and your data consist of questionnaire returns from several hundred visitors, you may want to describe the proportions that come from different locations, fall into different age groups or stay for different lengths of time. You might do this by producing tables or drawing diagrams such as pie graphs. Such data processing may be sufficient for your needs. Statistical analysis of your data can be confined to determination of numbers of observations that fall into particular groups (e.g. the number of National Park visitors that spend one night in the Park) and calculation of means, medians and modes (assuming your data are measured at least on an ordinal scale) if you want to get some idea of average values. If you wish to convey some impression of the spread of values and the variability in your data, then you will need to calculate such things as interquartile ranges, standard deviations and coefficients of variation.

A = B?

This question seeks to determine whether things are like each other. In fact, it's very difficult to address a question of this type using statistical methods. Statistical tests are based upon the idea that the result is unlikely to have occurred by chance. The alternative, or null hypothesis, is that the observed situation has indeed occurred by chance. Most statistical tests are based upon the notion of whether observed *differences* could have occurred by chance. Because of the need to bias the test in favour of the null hypothesis to avoid claiming there is a difference when, in fact, there is not, differences need to be large or frequent in order for the test to reject the null hypothesis. Conversely, the test will tolerate quite large differences while still claiming to see no significant difference between A and B. This means that the test is biased in favour of finding similarity, and if you are testing for similarity that's exactly the bias that you should be avoiding!

Let's take an example. Suppose you had come up with the idea that, contrary to the widely held belief, apples were just as likely to grow on orange trees as on apple trees. You measured ten apple trees and found apples growing on three of them. You measured ten orange trees and found apples growing on none of them. Analysis of your data using the χ^2 test gave a value for χ^2 of 3.53, which is insufficient to reject the null hypothesis at the 5 per cent probability level. Consequently, you can claim that your result supports your hypothesis that apples are equally likely to grow on orange trees as on apple trees.

You should try to avoid getting yourself into (i.e. design your research to avoid) situations where you want to perform this kind of test. That's not always possible. If you do end up wanting to test if A = B, then bear in mind the bias in the test you use.

A ≠ B?

Obviously, asking whether things are different is exactly the same type of question as asking whether they are the same. The difference is that this question is just what many statistical tests *are* designed to do. So you have a whole range of options available to you. Which test you need to use will depend largely upon the nature of the data that you have (see p. 72) and the type of difference you are looking for (see below).

Types of difference

There are three types of difference that you may be interested in where statistics can help. Is A different from B, on average? Does A have a different spread of values from B? Is A, in any way, different from B? Clearly, these three types of difference are not the same, and it's important that you are clear what type of difference you are looking for and the way in which such a difference relates to the question(s) you are investigating.

Is A different from B, on average?

This question is, effectively, one that deals with differences between means, medians and modes. The question may be restated as: 'Taking account of the range of values that A and B might take, are their means (medians or modes) significantly different?' Tests that are appropriate to this type of question are the z ratio and the t test (for data that are normally distributed) and the median test (for non-parametric ordinal data).

Does A have a different spread of values from B?

This question may be restated as: 'Do A and B have different probabilities of having the same value?' Tests that are appropriate to this type of question are the χ^2 test (for nominal data), the Mann-Whitney U test (for non-parametric ordinal data), the randomization test (for non-parametric interval data) and one-way analysis of variance (for data that are normally distributed).

Is A, in any way, different from B?

One way to approach this question is to test for the various types of difference in turn. So, for example, you could first undertake a median test and then carry out a Mann-Whitney U test to determine whether your two data sets have either different means or different spreads of values. Some statistical tests, however, are sensitive to any type of difference between two data sets. If, therefore, this is what you want to know then it may be appropriate to use one of these tests. Both the Kolmogorov–Smirnov test and the Wald–Wolfowitz runs test can be used on ordinal data to test for any kind of difference. Although these tests do have great usefulness in this respect, the result will tell you less about the nature of the difference between your data sets. In that regard, therefore, you might find them less useful than more narrowly focused tests.

In Box 7.3 we present a number of questions that might form part of a dissertation. Can you identify the type of difference that is being looked for in each case, and the appropriate statistical test?

Box 7.3 **What type of difference is being looked for here, and what is an appropriate statistical test for it? The correct answers are given at the end of the chapter.**

Question	Type of difference	Appropriate statistical test
Do these streams have different variability in discharge?		
Are the sizes of particles comprising these glacial deposits different?		
Do marriage distances in these villages increase with the size of the village?		
Are the gradients on east-facing valley sides steeper or shallower than those on west-facing valley sides?		

A > B?

This question is very similar to the previous one, in the sense that it is looking for differences. But here you are looking for a difference of a particular kind. In the case of A ≠ B?, a difference in any direction would have produced a positive answer. A could have been bigger, better, more widely dispersed, smaller, worse or more narrowly spread than B. So this question may be regarded as no more than a specific form of the previous one and all the same statistical tests can be used. There is, however, one crucial difference. This difference is expressed by statisticians as *two-tailed* and *one-tailed* tests. A ≠ B? is a two-tailed test, whereas A > B? is a one-tailed test.

All the statistical tests mentioned in the previous section will give as their outcome a probability for the null hypothesis that A = B. If this probability is sufficiently small (often taken as less than 5 per cent) then we will reject the null hypothesis that A = B and accept the hypothesis A ≠ B. This probability is based upon the size of the difference(s) between A and B to which the statistical test is sensitive. For example, imagine that you have chosen to study the effects of vegetation type on erosion of footpaths and that you have selected the width of the path as your measure of erosion. As part of your project, you have asked the question 'Does path erosion differ between grass and heather?'. This question can be refined to being an example of the first case given in the previous section, namely 'Is there a different amount of erosion on grass and heather, on average?'. Let us assume that you have measured a large number of paths on grass and on heather and that the difference in their average widths is such that the answer is 'Yes'. This answer could be reached in two ways, as shown in Figure 7.1(a). Either the grass paths are sufficiently wider than the heather paths or the heather paths are sufficiently wider than the grass paths for the *combined* probability of these two events to be less than 5 per cent. Since the two of them cannot be simultaneously true for your data, the probability that, say, the grass paths are wider than the heather paths needs to be less than 2.5 per cent for you to reject the null hypothesis.

Now consider a slightly different approach. From your study of grasses and heather you have come to the conclusion that heather is more resistant to trampling than is grass. You think, therefore, that routeing paths preferentially over heather might be a good way of minimizing footpath erosion. In order to test this hypothesis you measure widths of paths on grass and on heather. Provided that path widths on grass are sufficiently greater than those on heather, on average, you will be able to reject the null hypothesis of no difference between them. Again, you require a probability for the null hypothesis of less than 5 per cent in order to reject it. This probability can be obtained in only *one way* (as shown in Figure 7.1(b)). So the probability for this

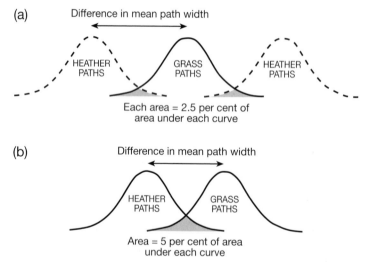

Figure 7.1 In (a) the null hypothesis of no difference between the widths of paths on heather and on grass can be rejected if the total shaded area is less than the critical value (commonly taken as 5 per cent). Because this area is made up of two sub-areas to allow for the possibility that path width on heather is greater or less than that on grass, a bigger difference in path widths is necessary than in (b) where the whole of the shaded area lies on one side of the curves. So, a bigger difference is necessary to reject the null hypothesis that path erosion does not differ between heather and grass than is required to reject the null hypothesis that path erosion is greater on grass than on heather.

single event, under the null hypothesis of no difference, can be as great as 5 per cent for you to reject the null hypothesis – that is twice as likely as when you were testing the possibility of either type of difference. As you can see, in Figure 7.1(b) the two curves showing the distribution of path widths can be closer together (so their mean values are closer) than either of the pairs of curves showing heather and grass paths in Figure 7.1(a). Consequently, the same difference in widths of paths on grass and heather may lead you to reject the null hypothesis of no difference if you were testing A > B? but to accept it if you were testing A ≠ B?. This example illustrates that an apparently quite subtle difference in the formulation of a question can have a major impact on the outcome of a statistical test. This is because the same test can often be used in many situations but the same result can have different meanings. Box 7.4 gives the outcomes of statistical tests. Do these outcomes allow you to reject (at the 5 per cent level) the hypothesis A ≠ B, A > B, both or neither?

Box 7.4 **What do the results of these statistical tests mean for the hypotheses? The correct answers are given at the end of the chapter**

Statistical test	Two-tailed probability level of the statistic	Hypothesis A	Accept/ reject the null hypothesis	Hypothesis B	Accept/ reject the null hypothesis
$\chi 2$ test	p = 0.06	The social mix of suburb A is greater than that of suburb B		The social mix of suburb A is less than that of suburb B	
t test	p = 0.03	The sizes of the particles comprising glacial deposit A are greater than those of deposit B		The sizes of the particles comprising glacial deposit A differ from those of deposit B	
t test	p = 0.11	Marriage distance is greater in large villages than in small villages		Marriage distance differs between large and small villages	
Mann– Whitney U test	p = 0.04	Male students obtain better degrees than female students		The quality of degree class differs between male and female students	

A ↔ B?

The search for relations among phenomena is one of the most important elements of research. It leads on to investigating why such relations might exist and to developing mechanisms for predicting one phenomenon from another. However, it is quite separate from such later steps and should not be confused with them. The simplest way to investigate whether A ↔ B is to draw a scatterplot, some examples of which are shown in Figure 7.2. Scatterplots not only provide simple, visual representations of relationships but can also be useful in trying to understand such relationships. Although many computer-graphics packages will readily draw scatterplots for you, it is often worth plotting them by hand (unless your data set is extremely large). Such plotting will often give you useful ideas about the nature of the relationships between phenomena. So, for example, you may find that subsets of your data show much closer relationships than the data as a whole, or that a wide scatter is due to a particular set of values collected at a particular time or in a particular location being quite different from the others. Checking outliers in a scatterplot against your field notes might reveal errors in data transcription (so that they turn out not to be outliers at all) or reasons why you might expect these values

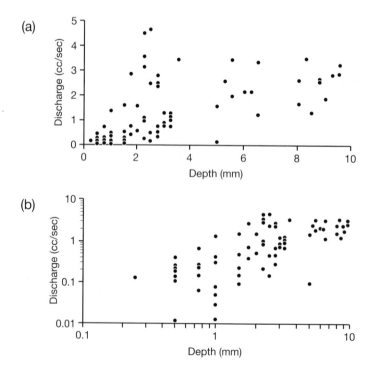

Figure 7.2 Scatterplots (in this case of runoff discharge against depth of flow) give a good visual representation of relationships between variables. Plot (a) shows that as depth becomes greater so the range of values of discharge increases and that as the values of each increase their relationship becomes more poorly defined. By taking logarithms of both depth and discharge, the relationship can be depicted as shown in Plot (b).

to be less reliable than the others (so you might consider dropping them from your analysis).

As well as displaying relationships between A and B, you may also want to quantify them. There are two forms of such quantification. The first is a measure of the strength of the relationship. Correlation coefficients are used for this purpose. The second is a mathematical description of the relationship, for which the structural relation can be used.

Correlation coefficients

Probably the most widely used correlation coefficient is Pearson's product–moment coefficient r. Formally, this statistic requires that both A and B are measured on at least an interval scale and that each is normally distributed. However, it is widely used even where these requirements are not met and computer simulations (Havlicek and Peterson 1977) show that even violent violations of these assumptions do not affect r very much.

For ordinal scale data there is a range of correlation statistics available that are computed from the rank order of A and B. Spearman's rank correlation coefficient ρ is the most widely used of these statistics. For data that are measured only on a nominal scale, the contingency coefficient C can be used. In many situations A and B will denote more than one variable (A_1, A_2, ... A_n and B_1, B_2, ... B_n) so that you may be interested in correlation coefficients between pairs among several variables. Such correlation coefficients are often presented in a correlation matrix (Table 7.2). Where several variables (say, A_1, A_2, ... A_n) are correlated with one other (say, B) and are also correlated with each other, the relationships among these variables may be quite complex. For example, assume you have been studying properties of grikes on a limestone pavement and that you have measured grike width, depth and length. Let us say that you have good positive correlations between grike length and width and between length and depth, but that width and depth are also strongly correlated. You want to know whether grikes are longer when they are deeper even if they all have the same width. One way you could try to solve this problem would be to go out and collect a set of data in which you measured only grikes that had the same width and then undertake a correlation of length and depth on these data. This might be difficult to do because the correlation between width and

Table 7.2 An example of a correlation matrix. This matrix shows the relationships between all pairs of parameters in a study of variation in rates of infiltration of rainfall into the soil at different locations. Note which pairs of parameters are positively correlated and which are negatively correlated.

	Rainfall rate (mm/hr)	A parameter	B parameter	Vegetation cover (%)	Fine soil cover (%)
A parameter	0.530				
B parameter	−0.249	−0.420			
Vegetation cover (%)	0.100	0.451	−0.116		
Fine soil cover (%)	−0.065	−0.146	0.017	−0.445	
Gravel cover (%)	−0.047	−0.367	0.111	−0.641	−0.402

depth may make it hard for you to find a good set of data in which depth and length varies over a wide range. The alternative approach is to determine the partial correlation coefficient between grike depth and length, controlling for width.

The use of partial correlation coefficients introduces the concept of *dependent* and *independent* variables. We will have more to say about those in the following section. Here, we can simply say that the dependent variable is the one that you are trying to find out about and independent variables are the ones between which and the dependent variable you are trying to find a relationship, and the ones whose influence on the dependent variable you are trying to control for. So, in the example, the dependent variable is grike length and the independent variables are grike depth and width. The partial correlation coefficient measures the relation between grike length and depth that is independent of the relation between grike length and width. Don't be surprised if the partial correlation between two variables turns out to be negative when their total correlation is positive. This can be the case if the correlation between the independent variables is stronger than that between the dependent variable and one of the independent ones.

Where your project involves several variables, many of which are correlated with each other, partial correlation can be extremely useful in sorting out how these variables are related to each other.

The structural relation

Not only may you want to measure the strength of the association between variables, but you may also want some way of quantifying the form of this relation in terms of an equation. Commonly, you will find regression techniques used for this purpose. However, often this is not correct. Strictly speaking, regression should be used where you want to predict one thing from another and there is a clear distinction between what you want to predict (the dependent variable) and what you want to use to predict it (the independent variable(s)). Where there is no clear distinction between your variables, and all you want is to describe the relationship between them, the structural relation is what you need to determine. This relation describes a line that lies closest to all the points in a scatterplot such that Σa_i is a minimum (Figure 7.3).

A → B?

If you ask the question whether A influences B, then there is a clear distinction between the dependent variable B and the independent variable(s) A whose relation(s) to B are being examined. You may be interested in the question whether changes in A are accompanied by changes in B or in assessing how well you can predict B from A. Regression techniques are appropriate to this type of question. Strictly speaking, regression should only be used in situations where you wish to be able to predict B from A. It is, therefore, wholly utilitarian. However, that does not stop you from using regression to answer the question 'Can you predict B from A?'. The fact that it turns out that it is possible might be interesting in itself, even though you don't want to do it. What is important is that you recognize that the ability to predict B from A says *absolutely nothing* about the causes of B.

84

Regression lines differ from the structural relation in that they have the property that either Σd_x or Σd_y is a minimum for the regressions of X on Y and Y on X, respectively (Figure 7.4). For the purpose of prediction this is an important property. It means that the regression line for Y on X provides the best estimate of Y given that you know X. Hence it is the best descriptor of the relation between X and Y only if you are sure you know X, that is, if you have measured X without error.

Regression techniques provide a powerful means of analysing your data. In particular, multiple regression, which allows you to look at the separate influences of many independent variables on a single dependent variable, can be extremely useful in sorting out influences on a phenomenon and in model-building (see Chapter 9). The following equation gives a multiple regression equation for predicting the A infiltration parameter (Table 7.2) from the rainfall rate, the vegetation cover and the cover of fine soil.

$$A = -0.374 + 0.0108 \; rainfall \; rate + 0.00721 \; vegetation \; cover$$
$$+ \; 0.00225 \; fine \; soil \; cover$$

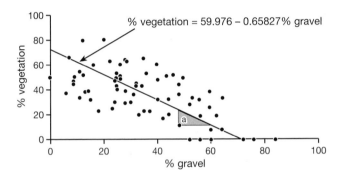

Figure 7.3 The structural relation describes the relationship between the percentages of ground covered by vegetation and gravel. The strength of this relationship can be obtained from the correlation matrix shown in Table 7.2.

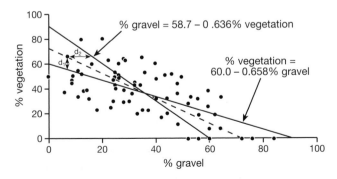

Figure 7.4 Regressions of vegetation cover on gravel cover (prediction of vegetation cover from gravel cover) and of gravel cover on vegetation cover (prediction of gravel cover from vegetation cover). Note the difference between the two lines and their relationship to the structural relation (dotted line) that was given in Figure 7.3.

Note that, although in Table 7.2 the correlation between A and the cover of fine soil was negative, in this equation there is a positive relation between the two. This is because of the stronger negative relationship between the covers of fine soil and vegetation (Table 7.2, p. 83).

A ⇒ B?

The question whether A causes B is probably the most interesting question of those listed in Boxes 4.4 (p. 38) and 7.2 (p. 76). Not surprisingly, therefore, it is the most difficult to answer and is the one for which statistical analysis is least helpful.

Consider the following example of research, published in *The Lancet* (Willett *et al.* 1993). In this study the authors are concerned with the possibility that the intake of partially hydrogenated vegetable oils increases the risk of coronary heart disease among women. The authors conducted a questionnaire survey of 85,095 women without diagnosed coronary heart disease to determine their intake of these vegetable oils and, subsequently, over a period of eight years monitored the incidence of coronary heart disease. They discovered that there is a statistically significant relationship between incidence of coronary heart disease and the intake of partially hydrogenated vegetable oils. The authors concluded that 'Our findings must add to concern that the practice of partially hydrogenating vegetable oils to produce solid fats may have . . . contributed to the occurrence of CHD [coronary heart disease]'.

Although Willett *et al.* (1993) stop short of claiming a causal link between intake of partially hydrogenated vegetable oils and coronary heart disease, they come quite close to such a claim. In fact, of course, their study provides no basis for establishing such a causal link. As the authors themselves note ' . . . it is possible that women with high intakes [of partially hydrogenated vegetable oils] were at increased risk of CHD for other unknown reasons'.

The concept of causation is not one that should be addressed at the stage of data analysis. It belongs at the stage of research design. Consequently, if your research question leads to the hypothesis that A ⇒ B, then this hypothesis will lead to certain predictions, such as A → B. It is such predictions that can be tested using statistical methods, thereby allowing you to falsify your hypothesis. That Willett *et al.* are unable to establish a causal link between intake of partially hydrogenated vegetable oils and coronary heart disease demonstrates the weakness of their research design, not of their data analysis.

Mistaking a correlation for a causal relation is one of the commonest errors to emerge from statistical analysis. Make sure you do not fall into this trap.

Qualitative data processing

In human geography, particularly, your data may be qualitative. They may come, for example, from interviews, observations, diaries or novels. In their raw form, your data are likely to consist of pages of text (transcripts of interviews, and so on). The task is to extract from the raw data something meaningful in the context of your research questions. Generally, this task involves some sort of coding of the data. You may do this manually by reading through the text and coding it as you go, or you may use a computer package, such as NUD-IST. Which approach you take is likely to depend on the amount of data you have. The benefit of a computer-based approach is that it will save you a lot of time if you have a very large data set. If your data set is small manual coding will have many advantages because of the greater familiarity with your data that it will give you.

Your coding should enable you to use your data to answer your research questions. Do not assume that because your data are qualitative your research questions are limited to A=?. The question 'Do young women of different ethnic backgrounds have different attitudes to community identity?' ($A \neq B$) can be answered equally well with qualitative data as it can with quantitative data. We would argue that all the types of question in Box 7.2 (p. 76) can be addressed using qualitative data, though, as with quantitative data, causation is the most difficult to address. Likewise, there is no reason why you should not use qualitative data in a hypothetico-deductive framework.

An important difference between qualitative and quantitative data is that for the latter there are agreed ways of doing things. So, if you state you used a χ^2 test on your quantitative data, everybody will know the rules of the test and you won't need to say any more, and you can jump straight to the result of the test. Because the 'rules' for analysis of qualitative data are not so well established you may need to give a more explicit account of your analysis before you present the result of your data processing.

Spatial data processing

Many research questions asked by geographers deal with spatial data. How does soil pH vary downslope? Do incidences of diseases vary with local economic inequality? Again, we think you can express any of these questions as one of the types in Box 7.2. The significance for data processing is that you may need to use different methods from those we've talked about so far if you want to examine some of the spatial properties of your data. For convenience, we can probably divide spatial data processing into geostatistics and GIS, though the two are not really exclusive, and you may well want to use both. In geostatistics the emphasis is on the statistical analysis of the spatial properties of your data: in GIS there is more emphasis on visualization of these spatial properties.

Chapter summary and conclusion

This chapter has shown how data processing fits into your research programme – specifically how it will help you answer your question(s). It has identified different types of data and examined appropriate ways in which you might analyse your data in order to use them to answer your question(s).

Suggestions for further reading

In this chapter, we haven't aimed to provide a potted course in data processing. What we have done is to indicate *types* of data processing that you need to consider in relation to your own research question(s). The reading you will need to do at this stage in your dissertation will depend on (a) what you are trying to find out and (b) your degree of competence in the methods you plan to use. However, the following may be helpful in a general way:

Clifford, N.J. and Valentine, G. (eds) (2003) *Key Methods in Geography*, Sage Publications: London, 572 pp.

Kitchen, R. and Tate, N.J. (2000) *Conducting Research into Human Geography*, Prentice Hall: Harlow, 330 pp.

Koch, G.S. Jr. and Link, R.F. (1970) *Statistical analysis of geological data*, Dover Publications: New York, 438 pp.

Limb, M. and Dwyer, C. (eds) (2001) *Qualitative Methodologies for Geographers*, Arnold: London, 303 pp.

Phillips, I.L. Jr. (1971) *How to think about statistics,* Freeman: New York, 201 pp.

Shaw, G. and Wheeler, D. (1994) *Statistical techniques in geographical analysis*, 2nd edn, David Fulton Publishers: London, 359 pp.

Siegal, S. (1956) *Nonparametric statistics for the behavioral sciences*, McGraw-Hill: New York, 312 pp.

What to do after reading Chapter 7

Consult books on methods of data processing to find out exactly which methods you will need to analyse your data and establish the data requirements of these methods (see also Chapter 8), i.e. make sure your question(s), data and processing methods all 'match'.

If you are reading this chapter before you've actually done very much about your dissertation (so don't yet know exactly what question(s) you are going to ask), then you might like to think about your question(s) in the light of the data processing you would like to do. If you are keen to do a project that will involve a lot of statistical analysis, then you should design your project accordingly. Likewise, if you are unhappy about using statistical techniques, don't start down a path that will inevitably lead you to them.

Answers to questions in boxes

Box 7.1   Ratio, nominal, ratio, ordinal, interval, nominal.

Box 7.3   A difference in the spread of data measured on an interval scale may use a t test; a difference of any sort for data measured on a ratio scale may use a t test; a difference in the spread of values for data measured on a nominal scale may use a χ^2 test; a difference in means for data measured on an interval scale may use a t test; a difference in means for data measured on a ratio scale may use a t test.

Box 7.4   Reject A and accept B; reject both A and B; accept both A and B; reject both A and B.

8

WHAT AMOUNT OF DATA
DO I NEED?

This chapter explains how the amount of data you collect will influence your ability to answer your question(s) and looks at how many data you will need to tackle the different types of question identified in Chapter 4.

Introduction

There are those who will tell you that you can't have too many data. They're wrong! The data you need are those that are required to solve the problem – *and no more*. Any more data means that you have been wasting your time somewhere and, given that the time you have is finite, losing time somewhere else. The big question is how many data do you need to solve your particular problem? The answer, of course, is that it depends on the problem.

Different amounts of data are needed for different things

The first thing to realize is that the size of your data set is only an issue if you wish to make inferences from your data. There are two reasons why you might want to make inferences. First, you may want to use the data you have collected in your study to draw wider conclusions.

Consider an example in which you are studying the business lifespan of shops that have ceased trading in a particular shopping complex and only ten shops have closed down since the shopping complex was opened. You feel that you have the time to study all ten of them. Your study will, therefore, tell you about these ten businesses, and, strictly speaking, no more. You may want to make some qualitative inferences from your study in which you suggest that your study may be representative of failed businesses in general. But, given the way in which you set up your study, namely that you deliberately chose to study a particular shopping complex, any such inferences would have no statistical validity. Whether or not this matters to you will depend partly on the question you are asking, and partly on the approach you have taken to answering it.

Now consider a similar project on a much bigger, older shopping complex. In this shopping complex 150 businesses have ceased trading since the complex opened. This number, you decide, is far too large for you to research them all. One approach would be for you to decide on a few specific businesses you want to study and for you to conduct your study on them. This project would be similar to the previous one. The result you obtain will apply only to those businesses you have studied and any inferences you may wish to draw from your study will be of a qualitative nature. There is nothing wrong with this approach, as long as you realize the limitations of what you can do with your results.

A different approach would be for you to decide that you want to study 150 businesses by taking a sample from which you will aim to make inferences about all 150. The question of how big this sample needs to be then arises. This chapter is about how you answer this question.

The amount of data you need depends on the type of problem you are tackling and the type of approach you are taking to the problem.

The second reason why you may want to make inferences is concerned with the nature of the question(s) you are asking. Even if you are in the situation of studying all ten shops that have failed in a particular shopping complex, you may need to think about the implications of the fact there are only ten such shops for the types of question you can ask. You might want to consider what factors may have led these shops to close down. Say you think that previous business experience of the owner may have been a factor and so you plan to compare the experience of the owners of your ten failed shops with that of the owners of the others that are still open. You may also think that the age of the owners is important so you look at that too. Then you realize that older people are likely to have more business experience than younger ones simply because they've had more time, so you plan to do a two-way analysis of variance. Now you will discover, when you come to check up on the appropriate statistical tests, that you have insufficient data for such an analysis.

Consider another example. In this case you are testing the hypothesis that, in a particular locality, east-facing hillslopes are steeper than west-facing ones. Now, if you measure one east-facing hillslope and one west-facing and it turns out to be true that the east-facing one is steeper, what does that tell you? Not a lot! You need more data. Discounting the possibility that they would have turned out to have exactly the same slope, one of them had to be steeper than the other. It could just as easily have been the west-facing one as the east-facing one. The number of measurements you need (amount of data or sample size) is a statistical problem – or more precisely a question of probability. It's the old coin-tossing problem where east-facing hillslopes are heads and west-facing hillslopes are tails. To identify bias in the coin, or to establish that east-facing hillslopes are steeper than west-facing ones, you have to get a result that is unlikely if there is no genuine bias. Statisticians usually accept a less than 5 per cent probability as necessary before the result is considered sufficiently unlikely to have occurred by chance. So, if you measure three pairs of hillslopes there's still a 12.5 per cent chance that you would find all the east-facing ones steeper even if there were no real difference. So you would have to measure at least five pairs of hillslopes *and* find all of the east-facing ones to be steeper to obtain a result that would support or refute your hypothesis. If one west-facing hillslope was steeper you'd need six out of seven steeper east-facing hillslopes for it to be deemed unlikely to have occurred by chance. Of course, if there is some geomorphological process that operates differentially to make east-facing hillslopes steeper than west-facing ones, it is unlikely to be the only process operating. Some other process affecting hillslope gradient may be making the west-facing hillslopes steeper or locally steepening hillslopes from time to time irrespective of their location. Thus, in the real world, measuring five pairs of hillslopes is unlikely to give a conclusive result. So, going back to the question of how many pairs you would need to measure, it becomes clear that the less important the process(es) causing east-facing hillslopes to be steeper, the more pairs you will need to measure to discover if it has any effect.

The bigger the effect something has, the fewer data you will need to identify it.

In the case of the coin, if it is very biased, it may come down heads every time, but if the bias is only slight it will come down heads only slightly more than 50 per cent of the time. If you are able to bet heads every time then a slight bias may be so small that your fellow gambler may not notice it, but it will be enough for you to have a certainty of coming out ahead in the long run. So with the hillslopes, if the process causing east-facing hillslopes to be steeper is only a very small contributory factor to their overall steepness, then you will need to measure a lot in order to discover if it is operating. This may sound depressing but, in fact, it can work to your advantage. If you discover the bias with a very small sample, it implies that the process making east-facing hillslopes steeper is very important in controlling their overall steepness. If it turns out not to be very important, you might ask whether it is worth investigating anyway! (But see the example given in Figure 8.4.)

These examples demonstrate that the amount of data you need depends on the type of problem you are tackling. In addition, it shows that for quantitative data analysis it can often be quite easy to specify the *minimum* amount of data that you *must* have, but the actual amount you need may be more difficult to specify. This means that you may be able to determine the minimum amount of data that are necessary to answer your question. The relationship of this essential minimum amount to that which is sufficient to do a really good project is less easy to determine.

In this chapter we will tackle these two issues. First, we will discuss the simpler issue of the minimum amount of data necessary to perform particular statistical tests. Second, we will deal with the issue of sufficient amounts of data for both quantitative and qualitative data analysis.

Necessary minimum amounts of data for statistical analyses

To look at this question, we will again use the succinct forms of question types that we introduced in Chapter 4 and used in Chapter 7. For each type of question it is possible to identify minimum amounts of data.

$A = ?$

The size of your data set is only of importance to a descriptive study if you wish to use statistics to describe a population from a sample. Otherwise, your data are as they are. You can define the mean of two observations just as validly as you can of two thousand. So, if you are studying the business lifespan of shops that have ceased trading in a particular shopping complex and only four shops have closed down since the shopping complex was opened, then you can describe the mean lifespan of those four. If, on the other hand, you wanted to know the age profile of shoppers in the same shopping complex, you would not be able to ascertain the ages of all shoppers who had ever visited the complex. Consequently, you would need to sample the shoppers. The question of the accuracy of your mean age then arises. If you interviewed only three shoppers, could you obtain as accurate an estimation of the mean age of all shoppers as if you had interviewed 300? The answer is very probably no, but may not be for your particular data set. The correct answer will vary. There are statistical tests to help you solve this problem. A useful method is that developed by Cox (1952).

$A = B?$

As we've already seen, statistical tests aren't properly used to address questions where you are asking if one thing is like another. Consequently, it is inappropriate to identify the necessary minimum sizes of data sets.

$A \neq B?$

As we have seen already, the question whether one thing is different from another can be broken down into a number of more specific questions about differences between A and B. Each statistical test that can be used to answer the various forms of this question will have a necessary minimum number of observations below which it is not possible to obtain a statistically significant result. Again, it is important to recognize that, as with $A = ?$, the issue of size of dataset is only relevant if you are assuming that your data sample a population.

To get a statistically significant answer to the question 'Is this coin biased?', you would need to toss it a minimum of six times.

If the question is 'Do these four west-facing valley sides have gradients different from these four east-facing ones?', then you can answer that question simply by comparing the four east-facing and the four west-facing valley sides. Such an exercise tells you nothing other than providing a description of the data you have collected. But this may be what you want to know. For example, suppose you were investigating the fact that the four east-facing hillslopes all had houses built on them and the four west-facing ones did not and you had come up with the hypothesis that this difference was due to the maximum gradient on which houses could be built. You could then falsify this hypothesis if you found that one of the east-facing hillslopes was steeper than one of the west-facing ones. On the other hand, if the question is 'Do west-facing hillslopes in this area have gradients different from east-facing ones?', then your measurement of four pairs represents a sample; and you need to find out whether it represents a sample that is too small for you to be able to answer this particular question. The number necessary can be determined using the binomial distribution.

Eric says: Where's that statistics textbook when you need it? If you are reading this chapter because you think you should be using statistics in your dissertation and Tony and Peter are using terms you don't understand:

Go and look them up!

A > B?

We've already seen an example of this type of problem in the introduction to this chapter. However, there we were only interested in one possibility, namely that the east-facing hillslopes are steeper than the west-facing ones. So, as we saw in Chapter 7 this is just a one-tailed version of A ≠ B.

A ↔ B?

Statistical tests of a relation between A and B test the hypothesis that the observed relation would have occurred by chance in samples drawn from the populations A and B if there were no relation between A and B. There are two aspects of such a relation in which you might be interested: its strength and its form. The strength of the relation between A and B is concerned with the value of r, the correlation coefficient, and it is measured using the t statistic.

The form of the relation will be given by the equation of the type: $B = c + mA$, where c defines the intercept on the vertical axis and m defines the slope of the relation (Figure 8.1). If you are interested in the form of the relation then both c and m are of concern.

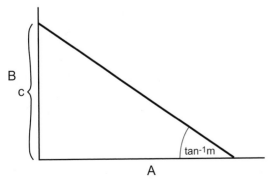

Figure 8.1 Graphical meaning of the values of c and m in the equation $B = c + mA$.

Tests of c

The most likely thing you will want to test about c is whether it is significantly different from zero. If you are studying the relationship between stream width and discharge, then there is a sound physical reason to expect that streams with zero discharge will also have zero width. So one thing you might like to test is whether your data make sense in physical terms such that you can assume c is zero and that the non-zero value that you have found lies within the range of sampling error.

Tests of m

The corresponding test for m, namely that it is non-zero, is, in effect, the test of the strength of the relationship because, under the null hypothesis of no relation-ship, m would be zero. The more interesting case, therefore, is to test whether m corresponds to some expected value. For example, you may have good theoretical reasons for expecting m to be equal to 1 so that changes in B are matched by changes in A that are of equal proportions.

$A \rightarrow B$?

Equations to predict B from A have the same form as those describing the relations between A and B. Strictly speaking, therefore, the equations can be defined using the same criteria for determining m. However, this is not generally done. For every independent variable A that is used to predict B, a minimum of ten pairs of obser-vations of A and B is generally regarded as necessary.

$A => B$?

As we saw in Chapter 7, simple statistical tests are not designed to investigate causa-tion. However, with good research design you can. The statistical tests that form part of this research design will address questions of the types covered in the previous sections.

How many data are enough to answer my question(s)?

The amount of data necessary to perform a statistical test is not likely to be the same as that you will require to answer the question you are investigating. Consider the example in which you are investigating the possibility that hillslope gradient affects pH of the soil. In order to perform a statistical test of such an effect you know that you need a minimum of three pairs of observations of hillslope gradient and soil pH. The result of such an investigation is shown in Figure 8.2(a). There is a relationship but it is considered to be too weak to be statistically significant (as is shown by the probability, under the null hypothesis of no difference, of obtaining the given t value). Now look at Figure 8.2(b). Your data set is larger. In this case (by chance), the value of the t statistic is the same but the probability of obtaining this value, under the null hypothesis of no effect, is less. This probability is still

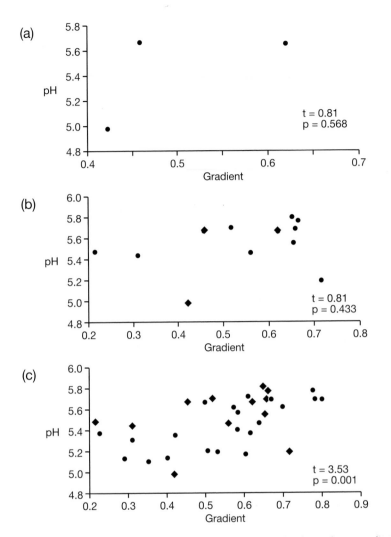

Figure 8.2 The outcome of a statistical test is strongly dependent on the amount of data you
have. Where relationships between variables are weak or subject to a lot of
scatter because of other influences, you need quite a large data set to obtain a
statistically significant result. Points that are common in 8.2 (a), (b) and (c) are
shown as ♦.

not sufficiently low for you to reject the null hypothesis of no effect. Now look
at Figure 8.2(c) which illustrates a case where your sample is even larger. The
probability of *t* is now less than 5 per cent and you can reject the null hypothesis.
So, it seems you can reach a different conclusion, depending on the size of your
data set. This may be true. However, what is also true is that if you don't have
the minimum amount of data you can't reach any conclusion. So, is it just a case
of getting past the minimum to perform the statistical test and then you have whatever conclusion you want? Not quite!

> If your coin always lands heads up, you could be reasonably certain
> that it is biased after you had tossed it six times. But if it landed
> heads up only three-quarters of the time, you would need to toss it
> twenty times to be reasonably certain of bias.

It is the case, however, that the bigger the sample the smaller the difference or the weaker the relationship between A and B needs to be for it to be statistically significant. But statistical significance is not an end in itself. It is merely a convention by which hypotheses are tested. Whether you should choose a sample size like that shown in Figure 8.2(a), (b) or (c) depends on what you want to know. A sample size that is close to the minimum for a statistical test would be appropriate for the question 'Does B have a very strong influence on A?', whereas a large sample size (such as that shown in Figure 8.3(c)) would be appropriate for the question 'Does B have any influence at all on A?'.

It might seem, at first sight, that the second of these questions isn't very important and that what you should be interested in is strong relationships and differences that can be identified from small samples. This will depend on the complexity of the problem that you are investigating. It might be the case that A isn't affected *very* strongly by anything at all, and this, in itself, is significant.

Figure 8.3 shows the relationships between distances moved by painted stones on a desert hillslope over a 16-year period studied by Abrahams *et al.* (1984). The results, showing relatively weak relationships between these distances and particle size, gradient and hillslope length, make it possible to discriminate among hypotheses regarding the mechanism causing their movement. If these particles were moved by gravitational forces alone then we would expect to find a strong relationship between the distance moved and the hillslope gradient (since this determines the downslope component of gravity) but no relationship between the distance moved and hillslope length or size of particle, because gravity is the same everywhere and, as Galileo showed, different masses are affected by it equally. If, on the other hand, the particles were moved by overland flow alone we would expect to find direct relationships between the distance moved and gradient and length of the hillslope and an inverse relationship between distance moved and particle size, because the amount of overland flow available to move the particles will be a function of hillslope length and its velocity will be a function of gradient, and smaller particles will be more readily entrained by the flow. That such relationships do exist allows us to reject the hypothesis of gravitational processes acting alone but the results are consistent with the hypothesis that the particles are moved by overland flow alone. However, the results are also consistent with the hypothesis that the particles are moved by a combination of the two processes. If this were so, and the gravitational processes were stronger, we would expect to find that the strongest relationship would be between distance moved and the hillslope gradient, since this factor affects both sets of processes but is the only factor affecting gravitational processes. As this is not the case, we can also reject the hypothesis that the particles move by a combination of processes in which gravitational processes are dominant. Thus we can conclude from the strengths of the three relationships that the particles are moved dominantly or wholly by over-

Figure 8.3 Relationships between distances moved by painted stones and (a) stone size, (b) hillslope gradient and (c) overland flow length. The strength of these relationships together with which of them is strongest makes it possible to discriminate among various hypotheses regarding the processes responsible for the movement of the stones (after Abrahams *et al.* 1984).

land flow. In this case, the fact that a particular relationship was weak and that others existed at all provided the information needed to test the hypotheses.

Quantities of data for qualitative analysis

As we have seen, the amount of data you must have for statistical analysis of quantitative data is determined by generally accepted rules. You can use fewer data than the rules specify (at least down to the point below which the statistics can't be calculated), and people often do, but if you do this, you need to be careful about the interpretation of your results. However, even for quantitative data analysis, knowing how many data you need to answer your question (as opposed to just performing statistical tests which may help you answer your question) is much less straightforward. For qualitative data analysis you have exactly the same problem. If you choose to analyse your qualitative data using statistical tests, then the same rules apply. But if you use a wholly qualitative approach to the analysis, then there are no clear rules about how many data you need. This fact makes life more difficult for you because you have to decide, and be able to defend, the amount (and nature) of the data you collect. There is a strong link between the amount of qualitative data you collect and the research question you can answer. Whereas with quantitative data you can plan the time to be allocated to your fieldwork knowing what is the minimum amount of data you must have and having a fair idea of a reasonable amount for the question, this is less true for qualitative data. While collecting your data you will need to be continually reflecting on the data and its relationship to your research question. For example, what you need to do after your first unstructured interview with a homeless person in Birmingham may well depend on what that person said during the interview, so you won't have been able to finalize all your planning in advance of starting the data collection. Reflection on your data as you acquire them may allow you to conclude the data collection sooner than you'd expected. (Have a holiday!)

Eric says: I polished off my data collection, and polished up my reflection, and it turns out that now I can go and work on my tan!

On the other hand, it may lead you to conclude that you need to do an extra week of work or that an answer to your question may not even be possible in the time you have available for data collection. Thinking about these things while collecting your data will avoid a nasty shock when you have your first meeting with your tutor after your fieldwork. In fact, if you go into that meeting ready to discuss such reflection, rather than with just a pile of scruffy notes you wrote up after your interviews, your tutor is likely to be impressed by the maturity of your approach.

Quality of quantitative and qualitative data

So far we've focused on how many data you need. But are all data the same? Do you need to think about the quality of your data in relation to the amount? We'd all like to think that the data we have (particularly those we've personally collected through hard work in the field) are wholly reliable. But, really, we all know that's not true. What about that measurement of stream depth you obtained hanging from the branch of a tree? Is it as good as the one you got standing on the bridge? Did you get the feeling that an interviewee was just making up a story for fun? If you know that some of your data are more dodgy than others, you should make notes at the time. You can use these notes during your data analysis. If some of your data seem a bit odd and your notes say that you had doubts at the time, you can use these notes to discard the dodgy data. Without these notes, you can't throw out some data just because they don't fit your ideas of how things should be. If all of your data are measured with less precision than you would have liked (if your shear vane is graduated in kPa but you need measurements at 0.1 kPa so you have to estimate the decimal point, for example), then this lack of precision introduces noise into your data.

Either because you have some dodgy data to throw out or because there's a higher level of noise in your data than you would like, you will need more data than you thought. So it's a good idea always to collect a bit extra than the bare minimum you need to answer your question, as insurance.

Chapter summary and conclusion

This chapter has examined the reasons why the amount of data you have matters. In particular, it has considered the amounts of data that are necessary to perform statistical tests and shown how this amount differs from the amount you may require to answer a particular type of question. This chapter should have helped you to determine the amount of data you will need to collect for your project.

What to do after reading Chapter 8

Look at Chapter 9 and decide whether modelling is for you. If you think it is, read Chapter 9. If you don't think that it has any relevance to your dissertation, then, having now read the chapters that deal with the data you will need for your dissertation, you should be able to decide:

1 what data you need
2 how you will collect them
3 how many data you will need
4 how you will process them.

To make sure that you can decide these things, make a list *now*! Check to see that the data will satisfy the needs of any statistical treatment you plan for your data. Talk to your tutors; get some advice. Don't actually *do* any of your data collection until you are happy about this list!

You may find it useful to try a practice run of your intended statistical tests on some mock data before you go into the field to check that it will work.

9

SHOULD I MODEL?

This chapter shows how modelling can be used as a method for finding things out. In addition, it looks at modelling in a wider sense and its role in dissertations.

Introduction

In the last few chapters, and particularly in Chapters 6 to 8, we have described a very empirical approach to solving problems. We have concentrated on an approach to problem solving that involves you in collecting and analysing data obtained from the real world. As we mentioned in Chapter 6, it may be that your institution will require you to take this approach to your dissertation. However, there are other approaches to problem solving that you might consider, other than straightforward statistical analysis of real data. In this chapter we will discuss these other approaches, as well as discussing more widely the role of modelling in your dissertation.

What do we mean by modelling?

Modelling is a term that is used very widely to cover a variety of things. In this chapter we want to concentrate on two ways in which modelling can contribute to your dissertation. First, modelling can be regarded as a technique for finding things out, much like fieldwork. You can use a modelling approach to actually do your dissertation. Second, there is conceptual modelling. A conceptual model is an abstraction of reality that expresses our ideas of how we think the world works. This type of modelling can contribute to the empirical type of dissertation we have been talking about so far, for example by presenting a summary of how what you've found out relates to its broader context.

Modelling as a technique for problem solving

Modelling is most useful as a technique for investigating problems that are concerned with processes. We can distinguish two types of modelling that you may employ in undertaking your dissertation: physical modelling and numerical modelling.

Physical modelling

Much physical modelling can be described as analogue modelling, in which you replace the process that you are interested in with something else which, you think, behaves in a similar fashion but which has benefits compared to the original process that make it easier to study. Many landscape processes operate so slowly or over such large areas that it is unrealistic to make direct measurement of these processes. An analogue process that behaves the same way, but which operates much faster or over a much smaller area is, therefore, potentially useful. For example, suppose you wanted to investigate how erosion processes fashion hillslopes. These processes take thousands of years to achieve their effect because the rock and soil of which the ground is composed are relatively resistant to the water, ice and wind, which fashion them. So you might decide to look for some more easily erodible material. One such material is ice. You could run water over a block of ice and see how the shape of the ice changed, or more usefully, see if it changed in the way you expected it to change.

Perhaps the most widely used physical models are flumes in which you can model river processes. Given the readily available data on river discharges, there are good opportunities to compare results from flume studies with real rivers and to use real data to guide you in designing flume experiments. Many physical geographers have specialist laboratories that would allow you to undertake physical modelling. For example, one of us has a low-temperature laboratory and the other has a rainfall simulation laboratory. Others have climate chambers in which you can simulate the processes of weathering. You may well have been introduced to specialist laboratories in your department during your course, but it's worth checking if other facilities exist in your department, or elsewhere in your university, to which you may be given access for your dissertation.

Although physical modelling does enable you to gain some insights into how processes work, it suffers from the limitation of how well the analogue retains the essential properties of the thing you are trying to model. Consequently, you always have to face the problem of deciding whether your results are telling you about the process you are investigating or about the differences between the analogue and the real thing. Nonetheless, this should not deter you from undertaking a dissertation that employs physical modelling, particularly if it is something you are likely to enjoy doing.

Numerical modelling

Numerical modelling provides a very powerful means of testing ideas about how the world works. Unfortunately, it is an approach that few students employ in their dissertations. This is probably due to the diffident way most students in geography seem to approach all things mathematical. However, if you don't suffer from such diffidence, undertaking a dissertation that employs numerical modelling is one way in which you might attract your examiners' interest (see Chapter 6).

Numerical modelling is a form of deductive logic (see Chapter 5). You define the rules (i.e. the processes) in terms of a set of equations and see what these rules produce. You can then compare the outcome of the rules with reality and, if the two don't match, you have shown conclusively that the rules are wrong. If they do match, then the rules may be correct. Much numerical modelling is undertaken by computer simulation so, as well as having a reasonable mathematical background, you are likely to need some ability at computer programming to undertake this type of dissertation.

To show how it can be used to investigate a problem, let's look at how Wainwright *et al.* (1995) used numerical modelling to investigate the formation of desert pavement. These authors came up with the idea that desert pavement might be formed by the action of raindrops falling on the ground surface. Obviously, there are different ways of investigating this hypothesis. They could have set up an experimental plot and seen what happened to the plot under natural rainfall. However, it doesn't rain a lot in deserts so they may have been waiting a long time for anything to happen and, what's more, the process may take a long time to achieve much. Second, they could have set up the same plot but used a rainfall simulator to solve the problem of the shortage of natural rain. Both of these approaches would have told them whether or not the action of raindrops did cause desert pavement to form. The third approach, numerical modelling, is a bit different. In the first two approaches they wouldn't need to know how raindrops work in order to test the hypothesis, but in the third they would. So the outcomes are slightly different. In the case of the first two (the field investigations) it is possible to discover whether or not the action of raindrops can produce desert pavement. In the case of the numerical modelling, if the simulation leads to the formation of pavement then this outcome *may* mean that (a) the action of raindrops can form desert pavement and (b) we know how the process works, *or* it may mean that their equations describing the process are wrong and that desert pavement will form under the operation of these incorrect equations but not under the operation of the correct ones. Conversely, if the simulation shows that desert pavement doesn't form, then it means *either* that the action of raindrops cannot form desert pavement *or* that the equations are wrong!

You may think this all sounds a bit pessimistic and inconclusive. You may be wondering why Wainwright *et al.* would have bothered with the modelling approach. One reason, obviously, is that sooner or later, we are going to need to understand how the processes work if we are to develop a true understanding of the formation of desert pavement (or anything else) and the modelling approach does provide a test of our present understanding. Furthermore, modelling, as we shall see, does provide useful insights that may not easily be obtainable any other way.

To run their model, Wainwright *et al.* needed suitable equations describing the processes of raindrop detachment and splash erosion. Like most modellers, they simply went to the library and found out what equations were available in the existing literature and used these. They also needed some initial surface to which to apply the processes described by their equations. Again, they went to the library and found some descriptions of desert soils. Consequently, they were, in this case, using secondary data for their modelling. With this information they were able to construct their model and produce simulations of the results. They found that desert pavement did form (sort of) but, interestingly, the nature of the pavement varied with the initial soil type. None of the existing literature on desert pavement talks of differences in the nature of the pavement in response to the initial soil type so this modelling study was able to identify something that might be worth looking for in the field.

Don't swamp your examiners with pages of undigested computer output.

Simulation modelling, of the type practised by Wainwright *et al.,* is a sort of computer game! Like most computer games, it is very easy to spend a lot of time on it. Whereas if you undertake field data collection for your dissertation you are likely to find yourself worrying about having enough data, such problems are unlikely to trouble you in simulation modelling. Change a parameter here, or an exponent there, and you can be off in a whole new set of simulations. The challenge is to know when to stop! Don't swamp your examiners with pages and pages of computer output. If you do, they will almost certainly ignore it and turn to the end of it to find out what you made of it. If the analysis and discussion of it aren't as weighty as the output itself, the examiners will probably take a dim view of what you have done.

Conceptual modelling

A conceptual model that expresses the way we think the world works can be useful in an empirical dissertation both early on, when you are trying to sort out the questions you are trying to answer, and towards the end, when you are trying to tie together the results of your empirical investigation.

Conceptual modelling and research design

A conceptual model can be used to show how the various parts of your planned study fit together.

To look at how conceptual modelling can assist you in designing your research, let's assume, for example, that you have decided that your dissertation will address the problem of the decline of retail services in rural Nottinghamshire. You might formulate a model for this decline as shown in Figure 9.1. Your model proposes that the decline in retail services is due to a falling demand for such services because of rural–urban migration and that this process provides a positive feedback to cause further decline in retail services. In addition, your model indicates a number of factors that contribute to cause the rural–urban migration. The conceptual model is here being used rather like a complex multipart hypothesis. This model, therefore, provides you with the questions you need to ask. The most important of these is 'Has there been rural–urban migration?'. If there is no evidence for this, then your model is false (see Chapter 5). If, on the other hand, you do find evidence of rural–urban migration, you might be prepared to accept your model as probably correct and focus upon the causes of the migration. If you investigate income, education and quality of life but find that none of them shows any difference between rural and urban areas, that would not invalidate your model. It would simply show that some other mechanism was responsible for the migration that you had observed. The conceptual model thus allows you to see (and explain) clearly how the various parts of your planned investigation fit together.

Figure 9.1 Proposed conceptual model for declining rural retail services.

Conceptual modelling and synthesis of results

As we mentioned in Chapter 2, one of the things your examiners will be looking for in your dissertation is evidence of intellectual achievement.

Producing a conceptual model out of the results of your empirical investigations is one way of demonstrating intellectual achievement in your dissertation.

Consider the example of the empirical investigation of the relationships between runoff and erosion rates and ground-surface characteristics conducted by Abrahams *et al.* (1988). These authors found correlations among runoff, erosion rates and ground-surface characteristics as shown in Table 9.1. From these correlations they produced the causal diagram shown in Figure 9.2 in which it is argued that surface roughness and surface-particle size jointly control infiltration, which, in turn, determines runoff. Runoff and gradient jointly control the erosion rate. Finally, both surface roughness and surface-particle size are related to each other and to gradient, but causal relationships among these three cannot be determined from the empirical observations. In this conceptual model Abrahams *et al.* sought to extend their empirical observations into a more general statement about runoff and erosion on desert hillslopes. This type of conceptual model is one way in which you can demonstrate the type of intellectual achievement that examiners will be looking for.

Table 9.1 Rank correlations of runoff and erosion rates with ground-surface characteristics (after Abrahams *et al.* 1988).

	Spearman rank correlation coefficient	
	Runoff coefficient	*Erosion rate*
Gradient	−0.77★	−0.31
Surface-particle size	−0.49	−0.14
Surface roughness	−0.66	−0.26

★ Denotes a statistically significant correlation.

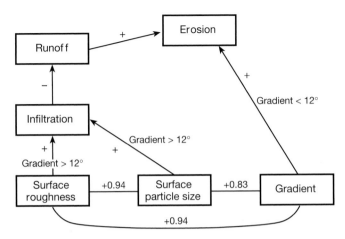

Figure 9.2 Conceptual model for controls of runoff and erosion on desert hillslopes (after Abrahams *et al.* 1988).

Chapter summary and conclusion

Modelling is a term that has a wide number of uses. In this chapter we have discussed modelling as a technique for solving problems, which can be employed as a method to do your dissertation, and conceptual modelling, which can be used to explain what you intend to do and what you have done.

Suggestions for further reading

Kirkby, M.J., Naden, P.S., Burt, T.P. and Butcher, D.P. (1993) *Computer Simulation in Physical Geography*, 2nd edn, John Wiley & Sons: Chichester, 180 pp.

Thomas, R.W. and Huggett, R.J. (1980) *Modelling in Geography*, Harper & Row: London, 338 pp.

Wainwright, J. and Mulligan, M.(eds) (2004) *Environmental Modelling*, John Wiley & Sons: Chichester, 408 pp.

What to do after reading Chapter 9

If you are reading through this chapter prior to starting your dissertation, think about taking a modelling approach to the problem you have decided to investigate. Check that such an approach would be acceptable under the rules of your institution.

If you are doing, or plan to do, an empirical study, look to see if expressing your problem as a conceptual model makes it easier to explain. If you haven't yet done the things listed at the end of Chapter 8, *do them now.*

10

HELP! IT'S ALL GONE HORRIBLY WRONG. WHAT CAN I DO?

This chapter tries to help you sort things out if they have gone wrong. It identifies the sorts of things that can go wrong at the various stages in your dissertation and suggests ways of making the best of the situation you find yourself in.

[STOP!]

[Have things really gone wrong?]

Many students think things have gone wrong when, in fact, they haven't. Before you read any more of this chapter make sure you are not in this category.

The commonest reason for thinking things have gone wrong (when, in fact, they haven't) is getting negative results. You may have set out with the hypothesis that economic wealth is strongly correlated with social status in urban populations, or that populations of limestone grikes are more effectively identified in terms of grike depth than width but then found that your results show no such thing. This may be disappointing for your plans to publish your dissertation in a leading journal, but it doesn't mean there's anything wrong with your dissertation. Remember, your examiners will be looking for evidence that you can design and carry out a piece of research, not that you have made a major discovery or that you happened to guess correctly in advance what your results would show. If you're not sure whether things have gone wrong or not, talk to your tutor.

Introduction: what can go wrong, and when?

All sorts of things can go wrong while you are doing a dissertation. Some problems can be so serious that they threaten the completion of the project, others threaten to spoil the quality of the work and hence to lower the mark that you will attain for your dissertation, and others simply make life difficult for you while you are doing your work.

The problems can strike at any time, and, inconveniently, problems often don't show up until long after you made the error or omission that caused them. It may seem harsh to be blaming you for your problems even at this early stage in the chapter, but it is fair of us to do so because even if you didn't directly cause your problem, you evidently didn't do enough to make sure it couldn't happen, or it wouldn't have! Maybe you have just discovered that the data you collected are inadequate for the statistical analysis you had planned. Maybe you are now standing at your field site and have discovered that the area contains none, or insufficient, of the things you had come to measure. Maybe you are back at your desk writing up your analysis and you have just discovered that your data aren't what you actually need to answer the question(s) you set out to answer. Maybe you've lost your disk and have no backup. Maybe you just don't think you can make the deadline.

If you are seriously reading this chapter because you have recognized that you have a problem, then you are already well ahead of the other guys who don't even know what a mess they're in.

Most of these problems can be prevented from arising by careful planning in the early stages of the project. Those that do arise can usually be dealt with, and dealing with them effectively can be the key to the success of your dissertation. What we will try to do in this chapter is to identify the main stages at which things can go wrong and what you can do about them.

Fixing the mess when things do go wrong

OK, so you are sure things have gone wrong; *don't panic*. Just because things have gone wrong there's no reason why your dissertation shouldn't still turn out well, if you can keep your head.

The first thing to do is work out where things went wrong. Then you can either go back and fix them, or do something else to undo or disguise the damage.

Problems at the planning stage

This is when the most serious problems are likely to occur. These problems are most dangerous because they are likely to be so fundamental that they undermine your entire project. What's more, you are unlikely to notice these problems at this stage so you could well be far into your project (completed your fieldwork and begun your data analysis) before you notice them. This is pretty serious, and the most fundamental of these problems could lead to disaster and a complete re-start. You must guard against them.

Eric says: If only you'd taken things more seriously at the planning stage and thought ahead more carefully, you could have avoided this.

We have put great stress in this book on the need for careful planning and a sound research design. It is because errors in planning or research design are hard to spot as you are making them that it is important that you give a lot of attention to these stages of your work. Planning your dissertation is all about choosing a good topic to investigate and designing a research programme that will lead to unequivocal answers to the questions you have asked. If you don't do these things then problems will be building up ready to strike you much later on. Don't say we didn't warn you.

Problems at the data collection stage

Problems can arise whether you are collecting your data in the field, in the laboratory or in an archive. Fieldwork commonly gives rise to more problems, or it appears to do so because you are likely to feel most vulnerable when in the field without much support. The good news is that these problems can usually be fixed. Some of the problems that may occur at this stage are covered in the following subsections.

Your field site is devoid of the features you hoped to study

One option might be to go somewhere else. But this is not usually an option that will be available, particularly if you have decided, as we have suggested in Chapter 6, to make your dissertation interesting by undertaking it in a foreign location. You may be restricted to that locality because that is where you can get cheap accommodation. Eric, who is going to be your field assistant, lives there; or that's where you have gone for your summer holiday. If you are stuck with the locality, look to see what features it does have. It must have something you can study. If you are in this situation, then you are back at Chapter 4! The thing to guard against here is rushing into a project without careful planning. Everything we said in Chapters 4 to 8 applies all over again. If you've got into a mess once, you don't want to make things worse by getting into another, different one because you're rushing through the important stages of planning. Most of the difficulties in doing a dissertation (or anything, come to that) lie in the unfamiliarity of the task. Once you know what you're doing, things get a lot easier. So, having planned a dissertation once, you should find it comparatively easy to do it the second time around! Try thinking back to some of the other topics you had briefly considered for your dissertation. Could one of them be resuscitated to help you now?

One problem you may well have is that you might not have ready access to a suitable library when you're away from your own institution. But, again, things might not be as bad as they seem. A bit of exploration and you could well find that there's more available than you expected. Even if there isn't a local university, a nearby agricultural college, for example, is likely to have many of the journals related to some aspects of physical geography or agricultural historical geography. Local government agencies may have material for urban or social geography research. Perhaps the material you need is available on the internet. Maybe Eric can email you something useful. If you want to avoid having to read a whole new literature, try to think of a new project that is closely related to the one you had originally planned so that all your earlier reading is not wasted.

Your equipment breaks

Once again, you have choices. One will be to try to recover the situation so that you can carry on as planned. You may try to get the equipment repaired or return to your department and get a replacement or to see if you can borrow something similar from a new source. If none of these options seems likely to work, then ask yourself if you really need the piece of equipment at all. Can you make do? For example, suppose you had planned to measure gradients along a stream channel using a level. Could you make do with an Abney level? The question you need to ask is, 'Will the reduced level of accuracy be such that my results are meaningless?'. If the answer is yes, then obviously you cannot make do with the lower quality data. If you can't repair or replace the equipment and you can't make do, then you will have to think of another project that you can do with the equipment that you do have. As in the previous case, if you are in this situation make sure you design your new project with as much care as you did the first one (see discussion above).

The events you planned to monitor didn't happen

Dissertations that are based on a need for things to happen can run into problems when they don't! Perhaps you planned to do a study of visitors to a particular picnic spot, but for the whole of the period that you had set aside to collect data it rained and nobody had a picnic. (Didn't you consider that possibility when you were planning? Tut tut!) Alternatively, you may have decided to look at the effect of discharge variation on pollution levels in a stream but throughout your period of fieldwork your stream had a remarkably steady discharge. One solution, of course, is to think of another topic. In effect, the problem is much the same as going to a field site and finding it devoid of the features you planned to study. Another solution would be to approach the problem through numerical modelling. You may be able to use a simulation study to come up with a model of how you think the phenomenon should behave. You could, for example, analyse the literature on picnics and develop a model to predict the numbers of picnickers who would visit your site under particular conditions or on certain days. Obviously, a dissertation that did this but then didn't provide any empirical testing of the model wouldn't be as good as one that did. Nevertheless, you could still do very well, particularly if your modelling led to a dissertation that demonstrated most of the attributes we listed in Chapter 2.

Problems at the stage of data analysis

The most likely problem for you to discover at this stage is that your data are inadequate. Despite what looked like a carefully thought-out data set, it turns out, now you come to it, that you can't actually answer your question with the data you now have and it's too late to collect any more. There are three reasons why you might be in this position.

First, you didn't think through the design of your research as well as you should have done. We hope that this book will have prevented this possibility.

Second, the data collection took even longer than you could have foreseen. Perhaps things kept going wrong in the field so that you started collecting data but then things happened so that the data had to be thrown away. For example, suppose you planned to measure variation in infiltration in response to differences in vegetation cover. To avoid complications you want each site to have a similar initial soil moisture but you need several days over which to make the measurements. After a few days it rains so you have to stop work and wait for soil moisture levels to drop back to the pre-rain value. As a result you obtain fewer samples than you had intended so the multivariate analysis you had planned won't now be valid.

Third, you have, in fact, changed your question from what you originally intended it to be. This may have appeared a very sensible thing to do. Once started on your data collection, you may have realized that there was more to the problem than you'd initially thought. As a result, you modified your questionnaire, but, unfortunately, didn't realize the implications this would have for your analysis.

Whatever the reason, you now find that your data don't match up with the question(s) you are trying to answer. So, what question(s) can your data be used to answer? Almost certainly, there will be some variant on the question(s) you wanted to answer that your data will be good for. Remember, your examiners will be looking for compatibility among the question(s), the data and the analysis. They don't need to know how you achieved that compatibility! It may be normal to start with the question(s) but you *can* start with the data and work backwards – and you may have to.

If this is the answer, what's the question?

At the end of the last section we implied that if your data can't be used to answer the question(s) you set out to answer, then you might be able to modify the questions to fit the data. This is very important. You can extend this idea a long way. Unless you have been unbelievably unfortunate or lazy, you will end up with some information about something.

We have stressed throughout this book the need for good research design and careful planning. But you want the best mark you can get for your dissertation and you may not be too fussy about how you get it. Hence the title of this section. Whatever data you have must address some question. Even the most serious of the fundamental problems that may have arisen from weaknesses in planning might be overcome at this late stage by thinking about what you can do with the data you have. You are unlikely to get a very good mark for your dissertation starting from this position, but the first 40 per cent is much easier to get than the last 40 per cent, so even an 'emergency rescue' dissertation will be worth something. Even in this position, if you can work out a logical sequence from your data back to a question, you might achieve a reasonably well-structured dissertation. The weakness will be that you won't have much choice in what the question turns out to be. If you find that your data condemn you to using a poor question, then that will affect the overall quality of your dissertation and its mark.

Facing up to the mess

If it turns out that you don't have very much to say in your dissertation because things have gone wrong, then you should consider how much of the mess you want to admit to. If you've managed successfully to work backwards from the data to a modest, if not great, question, you might want to present your dissertation as if nothing had gone wrong and offer it as a relatively minor study for which you might hope to get a lower second-class grade. On the other hand, you might think it is a better policy to admit to the problems you had in tackling a really interesting problem and present the work as a bold, innovative, but unfortunately flawed effort. Remember what we said about the reasons for doing a dissertation in the first place (Chapter 2). The primary purpose of your dissertation is to teach you how to go about finding things out and for you to demonstrate that you have, indeed, learned how to do this. Generally, you would expect to do this by achieving success, but a thorough understanding of why you failed might get you quite a long way. Being able to explain to your examiners what went wrong, why and what you would do about it if you had the chance to do your dissertation again can get you quite a lot of credit. You could include a whole section in the discussion or conclusion explaining how your problems affected the project, but only do this if you have intelligent things to say. Don't just put in a section of excuses. The examiner won't be interested unless you show what you have learned. Critical evaluation of your own work is another of the skills that undertaking a dissertation should give you. You will gain credit from your examiners if you demonstrate that you have acquired this skill.

Chapter summary and conclusion

This chapter has aimed to show that you can get out of trouble if problems have arisen with your dissertation. It has identified the sorts of problems that might arise and suggested ways in which you can deal with them.

What to do if you've had to read Chapter 10!

Don't panic. Keep calm. Sort out exactly what has gone wrong and see what remedies exist. If you need this chapter, then you probably also need to talk to your tutor.

11

HOW SHOULD I WRITE UP
MY DISSERTATION?

This chapter explains how to produce an effective research report, and describes what the structure and presentation of a good dissertation should be like.

Introduction: the research report

Your dissertation essentially involves two elements: the research project and the research report. The purpose of the report is to tell people about the research that you have done. The examiners' assessment of the dissertation will be based partly on the quality of the research and partly on the quality of the report. However, the examiners have to base their assessment of the research on what they see in the report; they don't have much else to go on! This means that you can make a mess of a perfectly good project by writing a duff report on it. However good the quality of your research, you rely on your report to communicate that quality to the reader. A weak or careless write-up could waste all the effort that you put into the earlier parts of the project. The aim of this chapter is to help you to make the very best of your project by producing the best report you possibly can.

The longer you give yourself to write the report, the better it is likely to be. As a general rule, it is worth trying to write up some of your dissertation as you go along. If your institution requires a literature review or research proposal to be submitted early in the dissertation process, you will be able to look back at this and use it as a starting point when you come to produce the final version of the early chapters of your dissertation. You will also find it useful to reflect on what you wrote back then in the light of your subsequent research. Writing up your methodology as you go along is also useful. It's much easier to write down what you *are* doing than to try to remember what you *did* six months ago.

The rules of your institution

As in previous sections of this book, we have to remind you first of all that your own institution probably has its own specific rules and regulations about the presentation of the dissertation. These probably include specific instructions about the length of the dissertation, the size of type and the line-spacing at which it is printed, and the way in which it should be bound for submission. The instructions might also be specific about the content of the report: how long the abstract should be, whether appendices are allowed, and what style of referencing is preferred, for example. Where your institution has specific guidelines, it is essential that you follow them. Be sure that you have received and read all the instructions that should have been issued. If you are uncertain about any of the instructions, seek advice from your tutor as soon as possible. Don't waste marks by failing to follow instructions. If the advice that we give in this chapter conflicts with the requirements of your institution, remember that it won't be us who mark your dissertation!

Typical examiner's comment: We'd love to give a higher mark, but the regulations are such that . . .

Structure and presentation

The purpose of your report is to communicate your project clearly to the reader. The essential elements of a good report are therefore the essential elements of clear communication: structure and presentation. The structure of the report is its organization; the way in which its various parts are put together. A sound structure involves a logical sequence of clearly defined parts. The presentation of the report includes both the overall 'look' of the work in terms of layout and neatness, and the clarity of the writing in terms of correct and effective use of English.

In any discipline, one element of the student's education is a familiarization with certain conventions that exist within the field. An architect, for example, must know the conventional symbols for use in drafting; a surveyor must know the conventions for writing down field measurements in such a way that a cartographer could understand and interpret them. In subjects where written reports are an important element of professional work, the student must know the conventions attached to the writing of reports. The reason that conventions exist is to facilitate communication. If we all expected different things of our reports and if we each tried to achieve our goals in completely different ways, we would find each other's work difficult to understand. Following the conventional styles is rather like speaking the standard language; it may seem restrictive in some ways, but it makes it possible for others to understand your work. Conventions exist in various areas of geographical report writing. Most obvious are conventions about referencing, writing abstracts and distinguishing observation from interpretation. In the following sections we will alert you to the most important conventions as we take you through each stage of the report. Of course, it is important to remember also that conventions can be ignored, if there is good reason. Our comments are intended as sound advice, not as binding rules.

Dissertation structure

To decide on the structure of your report you need to be sure of what it is trying to achieve. Any scientific presentation, be it a talk or a paper, should be made up of a small number of basic elements. There are many stories of learned professors standing up at the end of talks by younger colleagues to make comments along the lines of: 'My dear boy, I didn't understand a word you said; will you please tell me in just a few words what you did, why you did it, and what it means.' The point of the story is that a scientific report has some basic tasks to fulfil and that it should fulfil them as clearly and succinctly as possible. To explain your work you need to answer the following questions:

1   What is the aim of the work?
2   Why is this important?
3   How does it fit into a broader scientific context, including previously published work?
4   How do you set out to achieve your aim?
5   What observations do you make?
6   What do these observations tell you?
7   What, therefore, do you conclude?

These are the questions to which your report should provide answers. The list could be re-written like this:

1   What were you trying to find out?
2   Why?
3   How did you do it?
4   What did you find?
5   What does it mean?

You will notice that the questions follow on from one another in a sequence. The answers to each question in turn build up into a sort of story. The story goes like this: 'I set out to answer [this question], which is important [for these reasons]. I set about it [using these methods and this framework] and I made [these] observations. These observations indicate [this], so I conclude that the answer to my question is [this].'

The story runs like an argument, in a series of logical steps. Each step is clearly separated so that the reader can consider and judge each in turn, but the steps connect fluently together. In the previous paragraph we put bits of the story in square brackets. If you replace the content of the square brackets with your question, your methods etc., you will have your abstract. Try it! See also Box 11.7, p. 122.

Your dissertation should be a clear, logically constructed story.

The clarity of the story depends on its clear and logical structure. If you try to jumble up the components into different orders, or if you mix the different stages into one another, you will find that the story becomes more difficult to follow. You could try: 'I used these methods and the topic is important for these reasons. I reached this conclusion, which is an answer to this question. These were the observations I made.' At least the main components are there but the order is unhelpful, so no convincing argument is developed. If you miss out any of the key stages altogether, or if you add in too much spurious, unnecessary material, the situation deteriorates further. Chaos theory may have made its mark in the geographical literature, but chaos has no place in the structure of your dissertation.

The precise structure of your dissertation will depend to some extent on your particular project and approach. The structure of the report commonly mirrors (for better or worse!) the structure of the research on which it is based.

Research is a process of several stages, like a military campaign. Napoleon had to worry about mobilizing troops, assembling an armoury, attaining a strategic position for his flanking divisions, engaging the enemy at a distance and then moving in for close combat. You have had to worry about selecting topics and questions, assessing their wider significance, collecting background material, conducting a pilot study, collecting data, analysing data, relating your data to your question and reaching conclusions. One of the first tasks in drawing up a research design, or battle plan, was to assemble the elements of your research into a list and assign relative importance and proportion to each part of the list. We discussed this in Chapter 3. This enabled you to see the 'shape' of your project at a glance. Kennedy (1992) referred to the 'wine glass' and the 'splodge' as contrasting types of research plan, and the same distinction can be applied to different styles of report (Figure 11.1). The wine glass represents a well-structured project where the specific research being carried out is set into a wide context and its implications considered fully. The splodge is a poorly designed project where the research is given little context and where the wider implications of the research are largely ignored. Does your project look more like the elegant glassware or the dumpy splodge?

Eric says: **Don't be a dumpy splodge!**

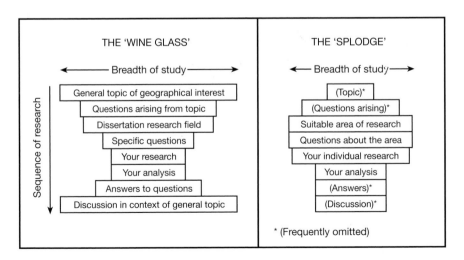

Figure 11.1 Two contrasting research strategies (after Kennedy 1992).

Although the character of your research will influence the structure of your report, some elements of sound structure are almost universally applicable, and we will emphasize these in the following sections.

Your report needs to be structured as a coherent argument, consistent with the conventions of geographical writing, so that you can communicate effectively with your reader.

Dissertation chapters

The main part of your dissertation needs to tell the sort of coherent story that we described in the section above. This is normally achieved by splitting the report into a series of chapters, each one dealing with one stage of the argument, or one of the questions in the lists above. Each chapter is commonly subdivided into a number of subsections, each dealing with a particular issue. A dissertation structured as a single, undivided unit would be unconventional to say the least.

The most common chapter headings include Introduction, Results, Discussion and Conclusion. Methodology, Scientific Background, Literature Review and Study Area are also common, but in many cases are relegated to sub-headings within the Introduction. As well as these central chapters, most institutions require an Abstract at the front of the dissertation, and you will also need to include a reference list and a list of contents. Box 11.1 shows the main elements of the typical dissertation or scientific report. You can look for other examples in the layout of papers in academic journals.

Box 11.1 The main elements of a dissertation or scientific report

1 Title page	8 Analysis of observations
2 Abstract	9 Discussion
3 List of contents/figures	10 Conclusion
4 Introduction (aims, background)	11 Reference list
5 Literature review	12 Appendices
6 Methodology	13 Acknowledgements
7 Observations/results	

In the following sections we will explain the purpose of the different parts of the report and offer advice on how best to produce them.

The title page

Most institutions have clear guidelines as to what your title page should include. In general, most institutions require that a page at the very front of your dissertation should contain:

- the title of the dissertation
- the name of the candidate (you)
- the date (year) of examination
- the examination (e.g. BA) and subject (e.g. geography).

You may be required to include:

- a formal declaration.

The form, and in some cases the exact wording, of the declaration will be stipulated by your institution. You may have to write it into the dissertation yourself, or you may have to fill in a form provided by your institution. Check the guidelines. The idea of the declaration is that you confirm that the dissertation is all your own work, and not somebody else's! A typical form for the declaration would be:

> 'Except where otherwise acknowledged, I certify that this dissertation represents my own unaided work.'

The title of the dissertation is important, as it is the label by which your work will henceforth always be known. Make it a good one. A good title should be accurate, clear, comprehensive and concise. You should not clutter the title with phrases like 'A dissertation on . . .' or 'A geographical analysis of', and you need to strike a balance between being too vague and too detailed. There may be a word or letter limit in your regulations (check this). Consider the examples in Box 11.2. Some typical layouts for effective title pages are shown in Figure 11.2.

Box 11.2 A selection of dissertation titles, good and otherwise

Shopping in Swindon	Too vague
A dissertation on shopping in Swindon	No better
Swindon as a regional shopping centre	Slightly better
Swindon's sphere of influence in regional shopping patterns	OK
A questionnaire analysis of Swindon's sphere of influence in regional shopping patterns	A bit too much?
A postal questionnaire study of Swindon's sphere of influence in Southern England and its effect on shopping patterns among young women (aged 18–35) in the region in mid-1996: a qualitative geographical analysis	Too much!

Figure 11.2 Some different ways of formatting a title page.

The contents page

First impressions are undeniably important, and the contents page is likely to be one of the first parts of the dissertation that a reader will look at. It is also a section that readers will refer back to over and over again as they progress through the dissertation. If they lose the thread of your argument, if they want to check back to a point you made earlier and need to locate it quickly, or if they want an overview of the structure and layout of the whole report, the contents list is where they will turn. A badly constructed or unhelpful contents list will spoil not only the first impression, but the overall impression that the reader has of your report. It is worth taking care to produce an informative and reader-friendly list of contents.

The purpose of the contents page is to list the various parts of the report. On one level it provides the examiner with a summary overview of your structure. On another level it helps the reader to locate particular sections of interest. It must be designed to fulfil both of these roles.

The most simple contents page is a straightforward list of chapters, as shown in Box 11.3. This is easy to construct and quick to read, but does not offer a great deal of detail to readers trying to find their way through your work. At the other extreme of detail, the contents page could include not only the chapter headings, but also each of the subheadings within chapters, and even the minor headings within subsections of the chapter. As shown in Box 11.4, such a comprehensive contents list could even include brief chapter summaries. The disadvantage of such a detailed list is that the structure of the report as a whole can easily become hidden in the morass of detail, so while the list is useful for locating specific items, it is less useful for showing the layout of the report.

The most effective contents lists fall between these two extremes, usually listing the chapter headings and the major subdivisions of each chapter, as shown in Box 11.5. There is much to be said in favour of restricting the list to a single page, so that the layout of the report is visible at a glance. Careful use of a hierarchical numbering system, subheadings of different orders printed in different styles (UPPER CASE, Underlining, etc.), and indentation can help the clarity of the page enormously. For example, compare the two lists shown in Box 11.6.

You may find that the word processing software that you use to produce your dissertation provides a facility for producing a contents list automatically from your text. It does this by recognizing different levels of heading and subheading in your text and using those as items in the contents. This can be very useful, but make sure that you impose your own rules on the contents list that is produced: don't let the software produce a microscopically detailed list if you had decided that a less detailed list would be better. Use the software but don't let it make your decisions for you!

The abstract

The abstract is a short section of text that summarizes the whole of your project. Just as abstract art aims to *abstract* the essence from its subject, so your abstract should *abstract* the essential elements of your work and report them in summary form. The regulations set out by your institution probably stipulate a maximum length for the abstract. Between 200 and 500 words is typical. The abstract is normally placed right at the beginning of a report, even before the contents list. When the reader opens up your report, the abstract is the first piece of text that appears.

The abstract serves several purposes. First, it enables the reader to find out what the project is all about before reading the whole report. This means that when he/she reads each of the following chapters the thread of the argument will be clear, and it will be easier to assimilate and to judge the material that is included. Second, the examiner will refer back to your abstract while reading the report if there is any confusion or inconsistency in the text. Also, having read the whole report the examiner can refer back to the abstract to check that the project has indeed done what the writer claims, and to check that the writer (you) has a clear grasp of the implications of the work. The abstract announces what you think are the salient points of your story, so the examiner can judge how well you appreciate your own work!

Box 11.3 Simple contents page

<u>Contents</u>

Box 11.4 A very detailed, rather unclear contents page

<u>Contents</u>

etc. . . .

Box 11.5 Part of a clearly set-out contents page

<u>Contents</u>

Box 11.6 Two different layouts for the same contents page, showing the importance of clear structure and presentation

Third, an abstract will be read by many people who will not read any other part of the report. This is true particularly of published reports; the abstract acts as a sort of trailer for the paper as a whole. For this reason it is particularly important that the abstract is entirely self-contained.

Given the prominence of your abstract at the front of your project and the fact that the examiner will read the abstract carefully several times, and also that some readers will base their opinion of your work almost entirely on the abstract, it is important that you write a good one.

A good abstract has all the elements of good structure that we described earlier for the report as a whole. Your abstract should explain, in a clear logical sequence: your aims; the background to those aims; your methods; your observations; your interpretation of those observations; the conclusions that you reach on the basis of those observations; and the implications of those conclusions. It should be a self-contained summary *not only of your results* but of your whole project. That is quite a lot to squeeze into a few hundred words, which is one of the reasons why good abstracts can be difficult to write.

One way of approaching the task is to write a series of one-sentence answers to the list of questions that we produced earlier when we were discussing structure. If you string the series of sentences together, you will have the basis of a sound, if perhaps rudimentary, abstract. If you try it, you will come up with something like Box 11.7.

Box 11.7   A stylized, simple 'model abstract'

The aim of this project is to (find out) . . . This is important because . . . This follows from earlier research on this topic in that . . . The research was carried out by (field/lab/documentary methods) . . . It was observed that/found that . . . This suggests . . . It is therefore concluded that . . .

You can increase the level of sophistication of this abstract by separating out the key sections of the structure into separate paragraphs. The first paragraph will be introductory and background material: the aim of the project, why it is important and the scientific background. The second paragraph might be the methodology: how you set out to do the research and the techniques you used. In some cases this might fit into the first paragraph as introductory material. The next paragraph should be your observations or measurements: the data that you collected yourself. As in the report itself, it is useful to keep the data separate from the background material and from the interpretation sections (we will discuss this more fully later in this chapter). Your final paragraph might cover the interpretation of your results, your conclusions and some comments, on the broader scientific implications of your findings (what it all means!).

This paragraph structure within the abstract should reflect the structure of your research. If your project dwells at length on methodological issues, then the abstract might reflect that with a paragraph devoted to methodology. If the methodology in your research is standard, then it might be absorbed into the introductory paragraph of the abstract.

Go to the journals for examples of how to write an abstract, but bear in mind that you will find bad examples as well as good, even in the published literature. The good ones will have a clear, effective structure. Study them, and you will see that they provide clear answers to the questions we have used to help our structure.

The introduction

The introduction to your report is not just 'the stuff you have to put at the beginning before you start'. It is an essential part of the logical argument that your dissertation puts forward. The purpose of the introduction is to explain to the reader what your research is about and what it aims to achieve, why it is important, what work has been done on the topic by previous researchers, precisely what you are trying to do or to find out and how you are going about it. These roles are indicated by the titles of the sub-sections into which the introduction is commonly divided:

- overall aim of project
- scientific background
- previous work/literature review
- specific objectives of project
- methodology.

In some cases, some of these sections can be presented as chapters in their own right. 'Literature Review' and 'Methodology' are frequently used as chapter headings rather than subheadings. Sometimes the scientific background and literature review are combined into a single section. The approach that you adopt for the organization of your introductory material will depend on several things, for example:

1   Institutional guidelines: some institutions require a literature review chapter.
2   Your material: you may or may not have enough to say on a topic to warrant a whole chapter, and you may or may not want to place so much emphasis on any of these sections in the context of the report as a whole.
3   The style of your argument: for example you may, or may not, wish to lead the reader gradually through the development of your ideas by separating the general background to your topic from the specific literature relating to your precise aims.

We will take each section in turn and consider what it should include, whether you produce it as a whole chapter or as part of a chapter.

Overall aim of the project

It is important to make it clear right at the outset what you are trying to do. It is very difficult for the reader to get involved in a report, or to follow an argument, if it is not clear where the story is leading. A bold, clear statement of aim will be a big help to your readers. The examiner will also be impressed to see that you can articulate your aims clearly. It is a common failing of undergraduate dissertations that the student does not seem to be very clear about what he/she is really trying to achieve. The 'Aims' section, as long as it is well written, can prevent this problem.

If in doubt what to write, begin with: 'The aim of this dissertation is to . . .'

At this point you should not be so specific in your aim that only an expert in the field will know what you are talking about. If this section comes right at the beginning, remember that the reader will not have seen your 'background' section and may be largely unfamiliar with your subject area. For example, if you write something like: 'The aim of this dissertation is to correlate the fabric of the upper and lower Twiddlecoombe tills with respect to the regional diagenetic glaciotectonic structures of the Late Devensian', the non-expert reader will not be very much wiser after they read the aim than before.

Likewise, you should not be too vague. Something along the lines of: 'The aim of this project is to study the tills of the Twiddlecoombe Valley' does little to convince the reader that you know what you are doing.

You need a statement that is broad enough to convince the non-expert reader, but specific enough to convince the expert. Consider the following example: 'The aim of this dissertation is to identify the extent to which subglacial till deformation occurred in the Twiddleton region during the Late Devensian glacial period by comparing the sedimentological characteristics of two glacial tills in the Twiddlecoombe valley.'

There is no single formula for the 'right way' to state your aims, but your method needs to be accurate, clear, precise and direct.

Scientific background, theoretical framework and justification

Having explained broadly what you are trying to do, you are ready to explain in more detail what the work is all about and why it is worth doing. Some people like to see the 'Background' section in advance of the 'Aims'. Others, including us, feel that it is easier to get the full benefit out of reading the background only if you already know what the aim of the work is.

The 'Background' section is there to fill the reader in about the details of the topic. For example, if you have just announced in your 'Aims' that you intend to examine the food supply/black market system in Harare, you need now to explain a little about urban food supply and black market economics in general, and perhaps about the special case of Harare in particular. If your aim was to make a geomorphological identification of a glacial limit on the Isle of Skye, you will need to explain something about the geomorphology and glacial history of the area. This section is your opportunity to demonstrate to the examiners your wide expertise in the field and your familiarity with a broad range of material that relates to your topic, as well as your chance to set your specific research into its wider academic context.

As we indicated in Chapter 5, it may or may not be necessary to provide a background to the theoretical framework in which your work is set. If you are taking a Marxist-feminist approach to food supplies in Harare, you will need to provide a justification for this approach and an indication of the particular insights that this approach will give you. Similarly, you may want to argue why a critical rationalist approach to the examination of the identification of the glacial limit on Skye is a particularly effective approach to the question you are asking.

From the background material should emerge the reason for your study: in other words the justification. Part of the justification will always be that we do not already know whatever it is that you are trying to find out. Thus an 'unknown' element should arise from your background discussion. You might argue that the location of the glacial limit remains unknown, for example, or that there is controversy about a particular element of the food supply system. This 'unknown' leads you directly and conveniently to the next stages of your argument. If something is unknown, finding it out seems like a sensible objective.

Specific objective of the project

When you wrote your 'Overall aims' section, you had to bear in mind that your reader was not necessarily familiar with your subject area, and so you avoided too much specific detail. If you place your 'Specific objectives' after your 'Background' you can be more confident that the reader has some knowledge; the knowledge that you have provided! This section therefore aims to explain in specific terms, with reference to the issues identified in the previous section, *exactly* what you aim to do. At this stage you must be very precise. It is easy to state your objective as one thing when what you have actually done is something subtly but distinctly different. That is a disaster; the examiner will chop your argument into little bits and send it back in a bin liner. Fear the viva.

A favourite viva question: 'What exactly was the aim of your research?' (If you have written one thing in your 'objectives' and done another in the project. The examiner cannot lose this game.)

It is often convenient to identify a number of specific objectives that build together to achieve the overall aim of the project. This can help you in the planning and execution of the research by breaking the overall aim into convenient chunks, and it also demonstrates to the reader the logical structure of your work. It also makes it possible to rescue the project relatively easily if bits of the research don't work as well as you hoped (see Chapter 10 for more details).

For the food supply/black market and glacier-limits examples that we used earlier, appropriate 'specific objectives' might be as follows (these are rather ambitious projects, by the way):

Title	Informal food-supply networks in Harare.
Overall aim	To identify food production and distribution structures in the informal economy in Harare, Zimbabwe.
Specific objectives	To map the distribution of urban agricultural sites in Harare, to identify landowners, food-growers and food consumers for each site and to identify food sources (formal and informal) for a sample of the urban population.
Title	Geomorphological evidence for the extent of Loch Lomond glaciation on the Isle of Skye.
Overall aim	To identify the maximum extent of glacier advance on the Isle of Skye during the Loch Lomond stadial.
Specific objectives	To identify the distribution of till, the locations of moraines and the limits of periglacial weathering features on the island, and to produce a geomorphological map on which glacial limits can be recognized.

The key points about both examples are that the specific objectives are indeed specific, that they are clear and that they contribute directly to the overall aim that was stated previously. (They are typical of many dissertation proposals also in that they are hopelessly ambitious, and in that achieving the objectives will not necessarily help to achieve the aims.)

Having read your specific objectives, the reader ought to know exactly what you are trying to achieve. This means, of course, that the reader will be in a strong position to judge what you do achieve!

Previous work/literature review

Having explained exactly what you are trying to do, your next job is to explain what work, if any, has been done on your topic by previous researchers.

If your topic is worthwhile, it is probable that some previous research will have been carried out on it. Your own work will, presumably, have been devised with these previous contributions in mind. The aim of your work should be to build on what knowledge already exists. It is necessary, therefore, to explain this existing knowledge and to mention the work that has led to it.

If your dissertation has both a literature review and a separate scientific background, the difference between the two should be that the scientific background provides the broad context of your work, whereas the literature review focuses specifically on previously published work and on previous literature that is directly related to your project. This might be work that has attempted exactly the same goals as your own, or has worked in the same field area, or used the same techniques.

If the aim of the scientific background was to give the readers enough information to be able to understand your 'Overall aims', then the aim of the literature review is to give the readers the information they need to see both your 'Overall aims' and your 'Specific objectives' in a detailed context.

Eric says: The literature review is also an opportunity to fatten up the reference list.

As we will discuss more fully in Chapter 12, one of the things that the examiners will be looking for in your dissertation is some evidence that you have become, to some extent, expert in your chosen field of study. One element of expertise is a detailed familiarity with and understanding of the published research in your field. One function of the literature review in the dissertation is to allow you to display this side of your expertise. You should take advice from your tutor as to how much emphasis your examiners will place on the literature review, but it is generally expected that you should demonstrate that you are very well read in your field. The exact length of your review will depend not only on the institutional requirements but also on your precise field of research. A thorough literature review in a broad field will be much longer than an equivalent review in a narrow field, and a review in a field where a great deal of work has been carried out will cover more items than a review of an obscure, little-researched field. There are also different expectations in different fields. In social sciences there is an expectation that the literature review will be extensive, both in the breadth of the subject-matter context in which your research is set, and in the exploration of the theoretical framework of your dissertation. In physical sciences, by contrast, the context is usually much narrower, and there is seldom an exploration of the theoretical framework. So, in the social sciences you might devote a whole chapter of your dissertation to a literature review, whereas in a dissertation in the physical sciences it is quite common to regard the scientific background and the literature review as the same thing.

Methodology

Having told the reader what you are trying to achieve, and why, you now need to explain how you are going to go about it. The purpose of the methodology section is to explain the procedures that you followed in carrying out the research. This explanation should be sufficiently clear and comprehensive that if somebody wished to repeat your work in the future, they could use your methodology section as an instruction manual as to how to proceed. In your dissertation, the methodology section also serves to convince the examiner that you really knew what you were doing and that you knew how to do it properly. Readers will also judge your results and conclusions on the basis of the quality of your methodology. If you've used a faulty or inappropriate method, your readers will be able to judge that your results are less useful than if you had used a better method. It is important that you acknowledge any weaknesses in your methodology. Your examiners will certainly spot them, and will penalize you if you appear not to have recognized them.

A favourite type of viva question: Can you explain why it was not necessary to use distilled water in the experiments? Your methodology chapter was unclear on this point.

The methodology section commonly covers the following issues.

Scientific approach

You may have talked a bit about your scientific approach in your background chapter, since your approach will be closely tied to the background from which your project arises. However, even if you think your approach is obvious, you should explain the philosophical basis and the procedural requirements of your approach formally within your methods chapter. For example, is your study quantitative or qualitative? Does it adopt a hypothesis-testing approach? What types of information are required to fulfil the project's aims and objectives? What are the characteristics of your research design? We dedicated a lot of space to research design in this book, and we hope you dedicated a lot of time and effort to research design in your project. This would be the place to explain your research design to your readers so that they are absolutely clear (a) how your research is designed and (b) that you understand how research design works. Writing about research design, of course, leads you directly into writing about the specific methods that you employed in your data collection and analysis (see below) so you should be able to make your methodology chapter into a fluent, coherent story.

Methods and problems of data collection

You should explain the precise details of your data collection procedures, including the measures taken to overcome problems that you encountered. Depending on whether your study involves field, laboratory or library research, this might require detailed description of experimental equipment, of field survey or sampling procedures, or of questionnaire design and implementation.

Methods and problems of data analysis

You should explain the reasons for selecting particular analytical techniques, the nature of the techniques chosen, the practical effectiveness of the techniques and any problems experienced in their application to your project. It is especially important to describe any modifications that you have made to standard procedures. If there are different ways of doing things, you will need to explain why you selected one way rather than another. It is often useful to refer back to previous work (see literature review and/or scientific background) and set your methods in the context of methods others have used.

The following examples show you some of the different sorts of material that can be included in the methodology section of different types of dissertation:

Example 1

Dissertation title
Swindon: local centre, regional sphere

Subheadings within methodology section
3.1   Retail outlet mapping
3.2   Questionnaire survey
    3.2.1   Design
    3.2.2   Implementation
3.3   Computer modelling

Example 2

Dissertation title
A geomorphological reconstruction of the Loch Lomond glaciation of the Isle of Skye

Subheadings within methodology section
3.1   Geomorphological mapping
    3.1.1   Use of air photos
    3.1.2   Field observations
    3.1.3   Feature recognition
        3.1.3.1   Glacial features
        3.1.3.2   Periglacial features
3.2   Till fabric analysis

The results section

Your dissertation should include a clearly identifiable section, be it a chapter or a series of chapters, which simply reports the data that were collected by means of the procedures previously described in the methodology section. This section must be separate from the previous sections, which explain the background to the project, and separate from the subsequent sections, which analyse and interpret the data. It is conventional practice for the data to stand alone in a section of their own. Future readers of your research can then distinguish easily between what actually happened in the field or in the lab and what you thought it meant. If your methodology was faithfully recorded and was satisfactory, future researchers might trust your observations even if they are not interested in your interpretation. If your results are clearly presented, the reader should be able to assess them and to evaluate the conclusions that you draw from them. The reader might choose to reach a different conclusion from the same results, but at least the results will have been identifiable and useful!

Results can be presented in a variety of ways, depending on the nature of the study and on the methodology used. A good way of getting ideas about how to present your data is to look through published work that uses similar types of data. See if you can adapt for your project any of the styles of presentation that you see in the literature.

Remember, it's usually a good idea to know how you will present the data even before you start to collect them, as the intended style of presentation might have implications for the style of collection.

The analysis of results

If you have collected data, they may well require some sort of analysis before you can effectively discuss what they mean. For example, if you have collected numerical data in an attempt to assess the impact of aspect on soil type at a range of different altitudes, you will probably need to carry out some statistical analysis on your data before you can tell what the data actually show. The data analysis has to be separate from the data presentation as we have already indicated, and the analysis has to come in advance of the discussion and conclusion (obviously) so a separate chapter may be appropriate. This will certainly be the case if the analysis is complex and extensive. For example, if you have collected data in the field on percentages of cover of particular plant species, but the focus of your research question is on the spatial characteristics of plant distributions you will need to devote quite a bit of your research effort into deriving the geostatistics that you will need before you can reach your conclusion. In this case a separate chapter will certainly be warranted. On the other hand, if your analysis is limited to simple statistical tests on your data you may be able to have a chapter that incorporates the analysis into the discussion.

The discussion

The purpose of the discussion is to bridge the logical gap between your observations and your conclusions. In this chapter you need to make very explicit the way in which the observations or measurements that you have made relate to the aims, background and scientific structure of your project. In your results chapter you said 'this is what I saw'. In your conclusion you will say 'this is the answer'. One of the main jobs of the discussion is to fill in the 'therefore'. You will also have to include and explain any 'notwithstanding', 'conditional upon' or 'level of confidence' messages that you wish to use to temper the final conclusion.

It is important to keep your discussion clearly separate from your results so that readers can distinguish between the supposedly 'objective' element of your measurements and observations and the more 'subjective' element of what you think about those observations. Again, a future researcher scouring the literature for data about your topic will be interested in your observations but probably not in what you thought about them.

There are many different ways of structuring this chapter. If your dissertation has involved several distinct elements, for example a fieldwork component and a laboratory experiment component, then the structure of the discussion might be determined by the need to discuss the two components separately. It may also be appropriate to include an assessment of any flaws or weaknesses that you recognize in your project. The following examples show a couple of different ways in which discussions can be organized:

Example A

Dissertation title   Informal food supply in Harare
Subheadings in discussion chapter   (a)   Limitations of data sources
                                         (b)   Synthesis of results
                                         (c)   Implications of results

Example B

Dissertation title   Extent of glaciation, Cader Idris
Subheadings in discussion chapter   (a)   Discussion of geomorphological data
                                         (b)   Discussion of palaeoecological data
                                         (c)   A model of former ice cover

The conclusion

The conclusion is the final stage of the logical argument that your dissertation has presented. It is the final position that your story arrives at. The conclusion is where you say: 'Given the evidence presented in the results chapter, and bearing in mind the issues considered in the discussion chapter, this is the answer to the question set up at the start of this dissertation.'

Your conclusion need not be very long, and indeed will do its job more effectively if it is concise and to the point. The main part of its job is to list the principal findings of your research. These can be divided into several types. First, and usually most important, are the points that answer the specific questions that your project was investigating. Second, there may be methodological or other findings that arose as a by-product of the principal line of research. Third, there may be issues concerning the nature of the research itself that warrant coverage as items of conclusion. Finally, you may have material to include under the general heading of 'Future research'; some institutions allocate specific credit to dissertations that identify avenues for the future development of the topic studied (see Chapter 12).

The conclusion is one of those sections, like the abstract, the contents list and the statement of aims, that the reader will refer to more than once. People who don't want to read the whole report might just look at your conclusion. The examiner will read it especially carefully. You need to be sure that the message of the conclusion is clear and easy for the reader to extract.

Unless the rules of your institution specifically recommend against it, it might be worth listing your conclusions in point form instead of using a conventional paragraph style. This makes it easy for the reader to see the major points without getting lost in superfluous wordage. The following examples illustrate two different ways of presenting conclusions succinctly and effectively:

Example A
Conclusion
The mapping of landforms on Cader Idris suggested that the corries were formed before the Loch Lomond readvance, and that the landforms showed only the last position of decaying glaciers, not the maximum extent of ice during the stadial period. The limits for the maximum extent of the advance were probably much further away . . . etc. . . . .

Example B
Conclusion
Three main conclusions arise from this study:
1   The corries in the Cader Idris range were formed before, not during, the Loch Lomond readvance.
2   The glacial deposits in the area do not reflect the maximum extent of glaciation, but a recessional position.
3   etc., etc. . . . .

Very often, institutions expect you to identify weaknesses and/or limitations to your study. Do not do this in the conclusion. The conclusion is *your* opportunity to say what *you* have found out. This is not the place to be mealy-mouthed and self-effacing. Certainly you do need, somewhere in your dissertation, to acknowledge that there are shortcomings to your methods, that your sample is rather limited or whatever. But there's a right and a wrong place to do it. The conclusion is the wrong place. The methodology or discussion chapter would be a more appropriate place.

Eric says: Go out with a bang, not a whimper.

The reference list

The purpose of the reference list is to provide full details of all the published material that you have mentioned in your report. A reference list is not the same as a bibliography. A bibliography is a list of publications that are relevant to your subject. A reference list contains only those items specifically mentioned in your report. These are items that you have 'referred to' in the sense of 'mentioned', not in the sense of 'looked at'. You should build up your reference list as you build up your text, adding references to the list as you mention them in the text. It is a difficult job to create a list only after you have written the text, and you are likely to make errors if you try to do it that way. One problem that can arise is that you might lose the details of an item that you used in the early stages of writing, and if you can't find the details you won't be able to use the item in your final text! Of course, you might have avoided this problem by using a reference-managing software package, but even without software a well-organized student should be able to keep track of all their references if they manage their work carefully. Manage your work carefully!

The idea of the reference list is that if your reader wants to go to the library and find a copy of something that you have mentioned in your report, your list will provide all the necessary publication details. Your reference list is therefore a reader service and an essential follow-up to the text.

In your dissertation, of course, the reference list serves another function. Inasmuch as your dissertation is an opportunity to demonstrate your expertise in the subject, the reference list is an opportunity to indicate how thorough you have been in your preparation for your research. When the examiner has a preliminary glance through the dissertation, a sick-looking list of less than a dozen references is usually taken as a reliable early warning that this is not going to be an expert piece of research.

We spoke earlier of the way in which well-known conventions in writing enable the reader to understand the writer. The reference list is a particularly good example of this. In the reference list you want to communicate a substantial amount of information – what the item is, who wrote it, when it was published, what journal it was in, etc. – in as compact but readable a form as possible, while being *certain* that the reader will be able to understand what you have written. To facilitate this there are a number of standard methods, or conventions, for listing references. The most commonly used in geography dissertations are the Harvard system and, to a lesser extent, the Numerical or Footnote system. Check your institution's regulations to see whether there are specific instructions about which method you should use.

The Harvard system

This system involves a very brief mention of (reference to) a publication in the text of the dissertation, and an alphabetical list at the end of the text giving full publication details of all the items referred to. The in-text reference includes only:

1   the surname(s) of the author(s)
2   the year of publication
3   the page(s) from which a direct quotation is taken (if applicable).

You do not need to include any other details such as the authors' initials or the title of the publication. Save those details for the reference list. The following examples show you what text references should look like:

Without direct quotation:

> ... several authors, including Smith (1982), Brown (1990) and Thompson and Jones (1992), have discussed the way . . .

> ... this has sometimes been considered a beneficial effect (e.g. Smith, 1982; Brown, 1990) and sometimes detrimental (Thompson and Jones, 1992).

With direct quotation:

> ... Smith (1982, p. 75) refers to 'natural order of chaos'.

> ... but the idea of 'a natural order of chaos' (Smith, 1982, p. 75) . . .

When you wish to refer to more than one publication written by a particular author in a particular year, differentiate between the items by labelling them 'a', 'b', etc.:

> ... as discussed by Smith (1982a) and Brown (1990). Smith (1982b) considers the problem of . . .

In the reference list at the end of the text, you need to include the full details of each item. These should be produced in the following form:

1   For a book
Author(s) surname(s)
Author(s) initial(s)
Year of publication (in brackets)
Title of book (italicized)
Name of publisher and place of publication

For example
Brown, A.B. (1990) *My Research in Chaos*, Big Book Co.: Littlehampton.

2   For a paper in a journal
Author(s) surnames(s)
Author(s) initials(s)
Year of publication (in brackets)
Title of paper (in quotation marks)
Name of journal (italicized)
Volume and issue number of journal
Page numbers of paper

For example
Brown, A.B. (1992) 'Experimental development of laboratory chaos', *Journal of Biochaotics*, 27 (2): 129–34.

3   For an article published in a book (e.g. in a collection of essays)
Author(s) surname(s)
Author(s) initial(s)

Year of publication (in brackets)
Title of article (in quotation marks)
'in'
Initial(s) of book editor(s)
Surname(s) of book editor(s)
Title of book (italicized)
Name of publisher and place of publication
Page numbers of article

For example

Thompson, A. and Jones, B. (1992) 'Recent progress in chaotic research', in P. Jackson (ed.) *Trends in Geography*, GeoPub.: Georgetown, pp. 276–82.

You don't need to put the publication date after the name of the editor, as inevitably it will be the same as the date you've given for the paper itself.

4   For a professional report

Author(s) surname(s)
Author(s) initials(s)
Year of publication (in brackets)
Title of report (in quotation marks)
Name of report series (italicized)
Report number, if any
Place of publication

For example

Brown, A. (1994) 'Ongoing research in chaotic states', *Institute of Chaotic Science, Report no. 94–8*, New York.

5   For a web page or internet site

Author(s) surname(s) (if available)
Author(s) initials(s) (if available)
Organization responsible for website (if no author is available)
The words 'on line' (in brackets)
Year of creation or last update, if available (in brackets)
Title of page or site
url (http address)
Date the site was accessed

For example

NASA (on line) (2004) *Earth Observatory*, http://earthobservatory.nasa.gov/ (accessed 20 September 2004).

The main point is that the reader should be able to trace your material, so, if in doubt, give more details rather than fewer in your reference. For other sources, such as newspapers, you can adapt the examples given above.

For more details and examples of this and other systems, refer to the reference lists in books and journals in your field. Many journals give explicit instructions in their 'guide to authors', often printed inside the cover.

Appendices

Include appendices only if you really need them. You will need them if you have material that needs to be included in the dissertation, but does not fit into any of the sections in the main body of the text. If you have a lot of data, such as transcripts of interviews, that the reader may need to access in order to confirm your analysis, you should not break up the text with pages of undigested data. Rather, you can put a summary in the text, and put all the data into an appendix. Make sure you refer to the appendix in the text, so that the examiner knows it is there and at what point he/she should think about looking at it. Appendices do not normally count towards the word length of your dissertation, so if you are running up against the word limit of your institution moving material into an appendix can be a useful tactic. Bear in mind that your examiner may *not* look at the appendices, so do not put anything there that is crucial to your argument.

Writing English

An essential element of your dissertation is communicating your work to someone who reads your report. An essential element of communicating your work to someone who reads your report is writing in a way that the reader will understand. You may or may not be specifically penalized for errors of spelling, punctuation, syntax or grammar (check your institution's guidelines), but such errors will make your report more difficult for the reader to understand, and might cause the examiner to misunderstand or misinterpret your meaning. This could affect your mark for better or worse. Probably worse.

Frequent student's whine: I thought this was supposed to be a geography exam, not an English exam.

Frequent examiner's laments: This looks like an upper second class project, but we can't give an upper second class grade to someone with this standard of English . . . . The research might have been OK, but I couldn't make much sense of the report, so I don't know.

Errors of spelling, punctuation, syntax or grammar will make your report more difficult for the examiner to read. This will irritate the examiner, and may make the examiner angry. An irritated or angry examiner is probably going to be a mean examiner.

Eric says: Don't irritate the examiner.

There is no room in this book to include a course in English, but it is important that you are able to write in a way that avoids irritating the examiner or making your report hard to read. If you have been told in the past that your English needs work, then it is worth giving it some. We talk a little more about how to write well in *How to do your Essays, Exams and Coursework in Geography and Related Disciplines* (Knight and Parsons, 2003).

Your dissertation should be written in a fairly formal style. This book is written quite informally and should not be used as a model! Avoid contractions (don't use 'don't'; 'isn't' ain't good either). Avoid colloquialisms (slang sucks).

Even if you think your English is OK, it is important to watch out for serious (and surprisingly common) errors such as the following:

Spelling mistakes
Punctuation errors
Sentences without verbs
Sentences beginning with 'But . . . ', 'And . . . ', etc . . .
Sipmle typnig errors.

Many serious and mark-damaging errors are made through simple carelessness. Students who can write perfectly good English sometimes submit work with the most appalling blunders. You should read through your work *many times* before you hand it in, checking for errors. Because you will be familiar with it, it will be difficult to avoid skipping ahead and missing small errors that the examiner, slowly and carefully working through the report, will stumble across. Reading it out loud can be a good way of spotting errors or poor expression.

Eric says: **Get family or friends to read the dissertation before you hand it in.**

Presenting text, figures and photographs

Your institution may have regulations about what you can and can't do. Check them before you follow our advice.

The main principle behind your presentation of the text and the figures in your report should be *clarity*. Your aim is to help the reader to understand what you are writing about. You can make a lot of progress just by writing good clear English, but the layout and appearance of text and figures are crucial. For a start, they are the 'body language' of the report; they help to signpost what is going on and where the reader is. Also, they are the medium of contact between you and the examiner. It's like getting a TV programme on a fuzzy old TV set – however good the programme might be, the duff reception ruins the effect.

Text

We have already discussed the layout of text in terms of dissertation structure, but text appearance is also important at a small scale. Each page should look neat. Each page should look like it is related to the other pages through a common format. Your institution probably insists on word-processed text, so neatness really should not be a problem. If for some reason you have somebody else do the word processing for you, make sure the person knows what you are trying to achieve and has read the institution's guidelines.

Eric says: **Text is talk. Speak clearly. Talk sense.**

Word processing is a very useful tool. It is easy to make everything look neat and tidy and it is easy to rearrange and correct text after you have typed it. This means that you should be able to produce *faultless* text, and that you can rearrange the work at the last minute in response to advice from anyone who reads your draft.

Caution: If you use a word processor spellchecker, be sure it is English, not American (unless you are submitting your dissertation to a US institution).

When preparing the text pay close attention to your institution's guidelines. These are usually very specific about:

- size of lettering
- width of margins
- line spacing
- weight (thickness) of paper.

Figures

Many figures, like your text, can be produced to a professional standard using the basic drawing packages that you probably have on your computer. It is worth taking the time to learn how to use one of these simple packages, even if it is only the drawing tool supplied within your word-processing software. If you don't have access to drawing software, you can achieve excellent results the old-fashioned way with ink, stencils and dry-transfer tone, but it's much harder.

Remember the key point: figures, like text, are a way of presenting information. Present only the information you *need* to present, and do so clearly and simply.

Some figures can be produced easily as part of word-processed text. Simple tables, for example, can just be typed into the text. Other figures can be produced via computer graphics in association with the data analysis stage of your work. For example, statistical analysis packages will allow you to produce hard copy of graphs and other material. If you use this approach, beware of three things:

1   Make sure that the quality of the output is equal to, or better than, the quality of the rest of your dissertation. It looks very bad if an otherwise nice report is presented in faint, smudgy, poorly aligned print on mis-matched paper.
2   Don't fall into the trap of including figures in your report just because the computer has given you figures. Examiners are on the look-out for undigested computer junk. Include only what your dissertation *requires* and *benefits from*.
3   Don't automatically accept the decisions made by the software about how to present the figure. For example, make your own choices about how to label axes, whether to box the outline of the figure, whether to label every data-point and so on. Your computer may have its own idea of what a graph should look like, but your computer doesn't actually know what your dissertation is trying to achieve. You do. So you should decide what does and doesn't go on your graph.

If you resort to drawing some or all of your figures by hand, remember that you do not need to be a professional draughtsman to produce a satisfactory illustration, as long as you keep it simple, go carefully and follow some basic rules. See if your institution has specific guidelines on this; most do. Personal advice, or at least a handout, should be available from the institution's drawing office. However, be aware that your hand-drawn diagrams will be marked alongside the ultra-smooth computer-drawn figures in your fellow students' dissertations. If you have pencil-drawn sketches with labels scribbled in ballpoint pen and bits of sticky paper glued on, the examiner will be unimpressed.

Remember, your dissertation is supposed to look like a professional piece of work, not a nursery-school project.

Photographs

You may wish to include photographs in your report, and it's usually a good idea. Check your institution's guidelines for any rules about this, but if there are no specific constraints from your institution, remember the same basic point that applied to other illustrations: clarity. Make sure that every photograph illustrates a specific point that is clearly identified in its caption and refers to a point that you've made in your text. Avoid including photographs that don't show anything. Do annotate and caption photographs to make the most of what they do show. Adding a clearly labelled sketch showing the key features from the photo can sometimes help to make your point clearer.

The best way to include photographs is to scan them to electronic files and then embed them within your text. If you use a digital camera to take the photos you can miss out the scanning stage, but remember to reduce your image file size to something appropriate. A dissertation with lots of 6MB images could become hard for your computer to deal with, especially if you try to print it on a central machine at your institution at the same time that 200 other students are trying to do the same thing.

Once you have electronic versions of your photos you can easily add labels or annotations to help make your information clearer. You can use the old-fashioned approach of gluing printed photographs onto the pages of your dissertation, but that isn't very smooth. If you can't scan your images, make an A4 colour photocopy of pages with photographs and bind the photocopy, rather than the original, into the dissertation. This prevents the unsightly bulges that arise when extra material is stuck into a bound volume. You can add labels on a transparent sheet bound into the dissertation as an extra page in such a way that it can be overlaid or lifted back to reveal the photograph.

There are lots of ways you can use photos, but the important thing is to use them effectively so that they help your dissertation and gain you marks, rather than simply loading your dissertation with unhelpful dead weight and losing you marks. Photos should serve a specific purpose, not just fill space.

Where to put the figures and photographs

Figures and photographs can be built into the text, placed at a point near to where they are discussed (every figure, photograph or table that you include in your disser-tation *must* be referred to in the text) or collected in groups, for example at the end of each chapter. The advantage of the latter is that the reader does not have to search through lots of pages to refer to a particular diagram, but the disadvan-tage is that the reader has to keep turning to the end of the chapter each time a figure is mentioned in the text. The approach you take will depend on whether most of your figures and photographs are mentioned only once in the text (put individually at the appropriate points) or are mentioned repeatedly in several parts of the report (put them in easy-to-find groups). If you embed photos at appropriate points throughout the text where they are first mentioned (the usual approach) but then need to refer back to a photo later in the dissertation, remember to refer to the page number, as well as the figure number, which will help the reader to find the photo easily.

Figure and photograph captions and their numbering

Every figure and photograph should have a caption; a short piece of text immedi-ately below the figure or photograph that describes what it is. The figure or photograph should also have a number, so that you can refer to it easily in the text. A good way of numbering figures or photographs is to number them sequentially within chapters. Thus the first two figures in Chapter 3 would be Figures 3.1 and 3.2. Tables in the text should be numbered separately. They can be numbered with roman numerals – Table I, II, III, IV, etc. – or they can be numbered sequentially within chapters, as for figures. Photographs can be numbered separately as 'plates' but this can lead to an overly complex system, and many people choose simply to include photographs among the figures.

Figure and photograph captions should be informative. Consider the following examples:

Example 1 Figure 3.1: Map.

Example 2 Figure 3.1: Map of the study site at Nether Wallop, showing the locations of the three sampling sites (A, B and C) and the area affected by the flood of April 1994 (shaded).

Remember, it can sometimes be a useful trick to simplify your figure by transferring information from the figure to the caption. On a densely crowded map, for example, the scale, or parts of the key, could be transferred to the caption.

How long should my dissertation be?

As usual, the first thing to do is check your institution's regulations. Most departments specify a maximum length that the report must not exceed. *Do not exceed this length.* Many departments also specify a minimum length. *Make sure your report is not shorter than this minimum.* Typically, the sort of length required is between about 6,000 and 12,000 words. Some institutions offer the dissertation as a single module, while others make it a double module. Check which your dissertation is supposed to be, as that will probably have a bearing on the expectations that the examiners will have of your work both in terms of length and intellectual achievement (and the implications of the mark you get for it for the class of degree you are awarded).

If your institution does not specify a length requirement, then you can work out roughly how long your dissertation should be by using the same criteria that the departments with length regulations use to work out their requirement. As we discussed earlier, your dissertation is supposed to do a number of different jobs: it describes your research; it discusses the background to the work; it considers the implications of your findings; it details your methodology and instrumentation; it demonstrates your expertise. It should do these jobs clearly but without undue padding. Depending on the scale of your project and the precise field of study this should take somewhere between 6,000 and 12,000 words.

Eric says: Size is not important.

Of course, Eric isn't strictly correct; size is important inasmuch as you must fit into the regulations and you will inevitably require a certain number of pages to do all the jobs you need to do. What Eric means is that 'bigger' does not necessarily mean 'better'. It is a common fault in dissertations to include superfluous material.

> **Typical student's strategy:** I've done all this analysis and it's turned out not to be relevant, but I'm not going to waste it so I'll put it in as an appendix.
>
> **Typical examiner's comment:** Bloody idiot, what's all this rubbish?

Binding and submission

Our advice here is very simple: have your dissertation bound in the style required by your institution and submit it on time.

If there are no specific guidelines as to how you should bind the report, there are many options available. At one extreme of convenience and (im)permanence you could use a simple ring-binder. The advantage is that it is quick, easy, cheap and amenable to last-minute changes. The disadvantages are that it looks unprofessional, can be cumbersome to carry and to read, and is not at all secure. More secure and more professional-looking binding techniques are preferable.

The important points to bear in mind when making your choice are as follows:

• regulations
• appearance, security and ease of handling
• cost and convenience.

Geography dissertations are prone to some particular binding problems, such as the incorporation of oversized maps or figures, and photographs. If photographs are stuck onto pages that are bound into a book, the extra thickness of the photograph can cause the book to bulge inelegantly. A good binder will overcome this problem by inserting spacers at the spine to accommodate the thickness of the photographs. The same problem can arise if you have foldout illustrations where one edge is bound at the spine and several layers are folded up between the pages. In such a case, be careful not to have the folded-over sections accidentally bound in so that the illustration is bound up in its folded and concealed state.

Chapter summary and conclusion

Your report tells the examiners about your research, and the examiners base your grade on what is in the report. You have to explain in the report what you aimed to do, why it was important, how you did it, what you found and what that means. You need to convince the examiners that you know what you are talking about and that you are 'expert' in your field of study.

There are numerous 'conventions' in the writing of scientific reports and you should adhere to these conventions so as to facilitate communication between yourself and the examiners. The structure and the style of your report are crucial to its success. There are many pitfalls to be avoided in your writing and many 'tricks' to effective presentation. A good report can make the most of a mediocre project; a bad report can ruin a good project.

You must follow the rules of your institution, produce your report accordingly and submit it at the right time.

Suggestions for further reading

There are many books offering advice and information on writing and presenting reports. Some that we have come across include:

Kitchen, R. and Tate, N.J. (2000) *Conducting Research into Human Geography*, Prentice Hall: Harlow, 330 pp.

M.H.R.A. (1996) *Modern Humanities Research Association Style Book*, 5th edn, M.H.R.A.: London, 100 pp.

Sides, C.H. (1992) *How to Write and Present Technical Information*, Cambridge University Press: Cambridge, 180 pp.

and, of course,

Knight, P.G. and Parsons, T. (2003) *How to do your Essays, Exams and Coursework in Geography and Related Disciplines*, Nelson Thornes: Cheltenham, 210pp.

What to do after reading Chapter 11

When you've read this chapter, you should have an idea of what your finished report should look like. If you have not yet written your report, then go ahead! If you have written your report, go through it and check whether it meets the general criteria that we have discussed; does it do all the things that a dissertation report should do? If not, do something about it. Remember, there is a lot of room for variety, but the general criteria that need to be met are quite standard.

Re-read your institution's guidelines periodically as you proceed to make sure that you follow all the rules. When you have finished, *don't forget to hand it in!*

12

HOW WILL IT BE MARKED?

This chapter describes examination and assessment procedures for dissertations, and explains what the examiners will be looking for. This includes both the written report and the viva voce examination, and there is some specific advice on how to handle the viva.

If you know what the examiners are looking for, it is easier to give them what they want.

Examination and assessment procedures

Each institution has its own assessment scheme, but the procedures for marking dissertations are pretty much standard throughout the system. In a typical scheme your project will initially be read and graded by a 'first marker' who will often be the project supervisor. The project is then read and graded independently by a 'second marker' who does not know what the first marker thought of the work. The second marker is usually someone in the department who has some specialist knowledge of the research area, but who may be less familiar with the material than the first marker. The marks assigned by the two markers are compared, and if they are similar a compromise mark, sometimes an arithmetic average of the two individual marks, will be agreed. If the two markers reach substantially different conclusions about the dissertation, and a compromise cannot be achieved, the dissertation will be sent to a 'third marker'. The third marker studies not only the dissertation but also the comments of the first two markers, and reaches a final decision about the work. The third marker is often an examiner from outside your own institution, and might decide that the only way to adjudicate your work is to interview you. In that case you will be called for a viva. We'll discuss vivas later in this chapter. Some institutions use an internal third marker, and only use the external examiner to adjudicate in exceptional cases. Other institutions routinely send many dissertations for external assessment.

There are exceptions to the general format of assessment described here. For example, some institutions allocate a portion of the marks available for the dissertation to elements of the work other than the final report. Some institutions allocate a small percentage to research proposals prepared in the first few months of the project, or to preliminary reports submitted immediately after the field season, for example. As with all the other aspects of the procedure, you need to know how *your* institution operates and how marks will be allocated to *your* work. Check with your supervisor if you are in any doubt.

What the examiners are looking for

In Chapter 2 we discussed the purpose of the dissertation and what constitutes a good dissertation. The issues that we covered in that discussion are the sort of thing that the examiner will have in mind when assessing your dissertation. You should keep looking back at that section while you are preparing your dissertation. Key points are:

1   A clear problem set in its scientific context.
2   A clearly explained and appropriate methodology.
3   Adequate and appropriate data and data analysis.
4   Clear separation of results and interpretation.
5   Sensible and penetrating discussion.
6   Logical and relevant conclusions.
7   Intellectual achievement and originality.
8   A high standard of presentation, easy to read.

Many institutions provide dissertation markers with clear guidelines in the form of a formal mark-sheet with specific questions and with space for examiners' comments. In some institutions the supervisor is given a particular set of questions about how the student went about the work, as well as questions about the finished report. Looking at the questions on these mark-sheets gives a very clear picture of exactly what the examiner is looking for. The examples in Box 12.1 are taken from the mark-sheets from two UK institutions, and give some insight to the examiners' way of thinking.

Notice the similarities between the points we discussed in Chapter 2 and the headings in Box 12.1. Bear these in mind when you are writing your dissertation. If you can get hold of a copy of the mark-sheet used in your own institution, so much the better.

A marking guide

Whether or not your institution uses formal questionnaire-style mark-sheets, there will certainly be a consensus as to what is required of dissertation students and as to what sort of mark should be allocated to dissertations of different quality. These standards should be constant not only between the individual examiners within each institution, but also between different institutions. One of the jobs of the external examiner is to ensure that this is so. Not every institution has formalized the grading of dissertations to the extent of producing a formal markers' guide, but such a guide is very useful not only to the examiners, but (more importantly) to you. Box 12.2 is a markers' guide of the sort that many institutions use. This guide tells you exactly what you have to do to reach whatever grade you want for your dissertation. Most

Box 12.1   Examples of headings on examiners' mark-sheets

Example 1

Selection of topic

Quality of literature review

Methodological overview and critique

Data collection

Data analysis

Results: presentation and
   interpretation

Conclusion (insight?)

Presentation

Overall impression

Recommended mark (per cent)

Example 2

(Specifically for an examiner who is
also a supervisor)

Description of project:

   conceptual difficulty

   technical complexity

conduct of project:

   independence

   originality

   organizational ability

   methodological awareness

   perseverence

   critical ability

   initiative

Written report:

   did you offer advice on writing?

   did you comment on a draft?

   academic content

   methods

   data handling

   presentation of results

   quality of results

   discussion

   suggestions for future work

   written expression

   evidence of plagiarism

Recommended mark (per cent)

institutions will have a qualitative marking guide that describes what is expected for a first-class, upper-second, etc. piece of work. This scheme is likely to be given to you in a handbook or published on your department's web pages. If you have been given a copy of such a scheme, you will recognize the guide in Box 12.2 as a version of the scheme, specifically tailored for dissertations.

First impressions

Your dissertation is a part of your final examination, and as such it will be assessed carefully and meticulously. Nevertheless, examiners are, believe it or not, human, and their approach to marking will reflect common human traits that you can exploit in your final presentation. For example, it is as true in dissertation marking as in anything else that first impressions are very important. Most examiners will glance through the dissertation quickly before settling down to read it thoroughly, and what they see on that first glance will colour their attitude to what they read subsequently. Typically, at the first glance, the examiner will:

- hold and look at the book as an object
- look at the title page

Box 12.2 A markers' guide for undergraduate dissertations

First class (70–100 per cent)
An excellent dissertation. Interesting research aims clearly set in the context of previous literature. Evidence of original and independent thinking. High quality reasoning and organization. Appropriate and clearly explained methodology. Sound and comprehensive data collection. Accurate and appropriate data analysis. Insightful and detailed discussion. Sound conclusions based on logic and data. High quality presentation.

Upper second class (60–9 per cent)
Falls short of first class on only a few criteria. May lack polish and fluency of a first-class dissertation, or may be flawed in some minor way. Still a good dissertation, meeting all the disssertation requirements, and meeting most of them at a high standard.

Lower second class (50–9 per cent)
Flawed in one or several areas, but nevertheless meeting the basic requirements. The level of detail, reasoning or presentation may be uneven. The evidence of insight, and breadth of reading, may be limited.

Third class (40–9 per cent)
Does no more than fulfil the basic requirements. Meets few of the criteria of a good dissertation. Reasoning, literature review and data may be weak or patchy. Presentation may be scruffy. There may be little evidence of originality or insight. The work may not be clearly set in a broader context.

Fail (0–39 per cent)
A broad category of marks to accommodate a range of dissertation types. These may include dissertations with evidence of plagiarism, dissertations that fail to meet the requirements of the institution, or dissertations that only meet the requirements at a most basic level. Different institutions will have different benchmarks within the fail category, but if you are reading this book before finishing your dissertation, you should not need to worry about them!

- read the abstract
- skip through the contents page
- flick through the pages of text
- look at the reference list
- maybe read the conclusion if it is short.

You need to make sure when you submit your dissertation, not only that your work will stand the close scrutiny of a thorough examination, but that the prominent parts of the work will do a good job of impressing the examiner on first acquaintance. If some of your photographs fall out when the examiner picks up your dissertation, it is obvious at first glance that your contents page is incomplete, or if your text appears to have been printed with a faulty print cartridge, the examiner will inevitably form a poor first impression.

Having read the dissertation, for a reminder of key points and to help finalize a mark, the examiner will probably glance again at the key elements. Commonly, having studied the dissertation carefully, the examiner will finally:

- re-read the abstract
- re-read the conclusions
- re-read the contents page.

That will usually be enough to check that the work is well planned and clearly structured, and to judge the extent to which it has done the job it set out to do.

Before you submit your dissertation, go through the same routine as the examiner and try to see your dissertation as the examiner will see it. If pages fall out when you flick through the dissertation, do something about it! Ask yourself the questions the examiner will ask. Give your dissertation a mark. Ask yourself what you could do, even if it is a last-minute job, to improve your dissertation. Better still, get a friend to 'mock' examine the dissertation for you. The best time to realize your weaknesses is before, not after, the formal examination.

Plagiarism

When we wrote our first draft of this book and showed it to colleagues in other institutions to get suggestions as to how it might be improved (just as you should do with early drafts of your dissertation), it was suggested to us that we should be more forceful on the issue of plagiarism. The strength with which this message came back to us is a measure of how strongly our colleagues – *the people who will be marking your dissertations* – feel about it, so take note.

Plagiarism is unacknowledged copying. At one extreme it can involve copying a whole document (a government report, for example) and claiming it as your own work. Less extreme, but no less serious, cases could involve copying sections of reports, articles, maps, books, web pages or other people's dissertations. Even using short passages, or ideas, from other people's work without specifically acknowledging the source, is plagiarism. If you lift anything from anywhere, you must acknowledge the source. You can acknowledge it by means of a conventional reference and/or in a separate section of acknowledgements. If you are deemed to be guilty of plagiarism, there could be serious legal implications, and, of course, the examiners will be singularly unimpressed. Your institution might simply disallow the whole dissertation, so watch out. If in doubt, consult your tutor for guidance. Your institution will almost certainly have guidelines on penalties for plagiarism. These guidelines will apply to your dissertation. Do not be fooled into thinking that your dissertation is so specific that nobody will recognize the sources from which you have lifted material. Your examiners are a very experienced bunch of people! If one of them gets suspicious, he/she will ask around and, if material has been plagiarized, it's almost certain that somebody will know the source.

The viva

Lots of students dread the viva and pray that they are not called for one. There really is no need to worry about the viva, though, because it will not be used to reduce the mark you have already been allocated on the basis of your written report; it will only be used to raise your mark.

In giving you a viva, the examiners are giving you an opportunity to improve your score.

You could be called for a viva for any one of several reasons. The viva might relate to your whole spread of examinations, not just the dissertation, and it might relate to an individual paper other than your dissertation. In many institutions vivas are given only to students whose marks are just below the borderline between degree classes, so that the examiners can judge whether the student deserves to be raised into the higher class. Some institutions viva all students.

The viva is like a cross between a tutorial and an interview. There may be one or several examiners, and they may want to talk with you about your exam papers, or your dissertation, or your coursework. Remember, the examiners want you to do well. They are trying to give you the opportunity to score marks. They are not trying to trick you or trap you into losing marks. If you relax, answer the questions that the examiners ask and try to engage in some discussion with the examiners, you will do well.

Remember the list of things that the examiners are looking for and try to give them exactly those things. You need to come across as a keen, interested student who has worked hard and who wants to do well. You need to demonstrate your ability to recall information from your years of study, to think on the spot about new questions that the examiners will throw at you and to argue logically about issues with which you are familiar, such as the design of your dissertation.

You do not need to defend your dissertation to the death if you know that it was in some way flawed. The examiners will be much happier with a student who sees the weaknesses of his/her work and can see ways of improving it if given another chance than with a student who thinks his/her work was just fine and cannnot be improved. There is always room for improvement.

Go into the viva prepared to explain how you'd do it better if you could do it again.

Honesty is always a good policy. If the examiners ask a question that you can't answer, don't pretend to know more than you do. Fumbling around foolishly trying to bluff your way out of a tight spot looks comical from where the examiner is sitting. The best approach is to admit that you don't know, but then to offer some ideas about how you could work out the answer or what you could do to improve your dissertation, depending on what the answer turned out to be. Remember, the examiner isn't just testing what you know, but how you can think and discuss. If you are called for a viva, follow the advice in Box 12.3.

Box 12.3 How to prepare for your viva

To prepare for your viva:

1   Be sure that you are familiar with the work that you have done. Read through a copy of your dissertation and get someone to ask you questions about it.
2   Think about how you could do the dissertation better if you had the chance to do it again. Look back through this book and remind yourself of what you should have done!
3   Talk to your tutor or supervisor and see if they have any advice to offer.
4   Check the library to see if any new work in your field has been published since you wrote the dissertation. The examiner will be very impressed if you demonstrate that you have kept up to date with the literature (especially if he/she has written any of it!)
5   Have a few early nights before the viva!

Chapter summary and conclusion

Your dissertation will be marked independently by several markers, some of whom may be experts in the field and all of whom will be concentrating on the specific criteria by which your dissertation is to be judged. These criteria are embodied in the markers' guides in Boxes 12.1 and 12.2. You need to be aware of these criteria and to make sure that your dissertation meets them. If you are called for a *viva voce* examination, remember that it is an opportunity to improve your mark.

What to do after reading Chapter 12

If you have not yet finished (handed in) your dissertation, then 'examine' it yourself using the criteria we've discussed in this chapter. Make any improvements you can before you hand it in. Remember to keep a copy for yourself that you can use to prepare for a viva if you have one. If you have been called for a viva, follow the advice in Box 12.3.

13

A FINAL WORD

Dear Reader

Is your dissertation complete? If you're all finished then it's nearly time to say goodbye. We wrote the first edition of this book because we believed that your dissertation was the single most important piece of work that you would do in your degree programme, that you needed all the help you could get and that we might be able to help. When we were asked to produce a second edition, we decided that you still needed help and that we could still offer it, so we did.

We think that research is an exciting and fun thing to do, and we hope that the way we have presented this book hasn't taken the fun out of it for you. We also think that research is largely based on common sense. We hope that this book has not made it seem more complicated than it needs to be.

Successfully completing a dissertation should give you a lot of things: a module contributing to your degree, of course, but more than that the confidence to know that you can identify a problem, devise a research strategy, carry out a project and produce a final report. Those are skills that you may well need again long after your dissertation has been consigned to the filing cabinet of history (or, in our case, geography). We hope this book has helped you to grow in confidence, as well as in competence. We hope that it has helped you to enjoy doing your dissertation, and we also hope that it has helped you to produce the best dissertation that you could.

Good luck with your next research project.

Tony and Peter

REFERENCES

Abrahams, A.D., Parsons, A.J., Cooke, R.U. and Reeves, R.W. (1984) 'Stone movement on hillslopes in the Mojave Desert, California: a 16-year record', *Earth Surface Processes and Landforms,* 9: 365–70.

Abrahams, A.D., Parsons, A.J. and Luk, S.-H. (1988) 'Hydrologic and sediment responses to simulated rainfall on desert hillslopes in southern Arizona', *Catena,* 15: 103–17.

Best, S. (2003) *A Beginner's Guide to Social Theory,* Sage Publications: London.

Clayton, K. (1981) 'Explanatory description of the landforms of the Malham area', *Field Studies,* 5: 389–423.

Clifford, N.J. and Valentine, G. (eds) (2003) *Key Methods in Geography,* Sage Publications: London.

Cox, D.R. (1952) 'Estimation by double sampling', *Biometrika,* 39: 217–27.

Dunkerley, D.L. (2003) 'Determining friction coefficients for interrill flows: the significance of flow filaments and backwater effects', *Earth Surface Processes and Landforms,* 28: 475–91.

Feynman, R.P. (1985) *Surely You're Joking, Mr. Feynman!,* W.W. Norton: New York.

Gleick, J. (1992) *Genius,* Little, Brown & Co.: London.

Goodman, N. (1978) *Ways of Worldmaking,* Hackett: Indianapolis.

Harris, C., Davies, M.C.R. and Rea, B.R. (2003) 'Gelifluction: viscous flow or plastic creep?', *Earth Surface Processes and Landforms,* 28: 1289–301.

Havlicek, L.L. and Peterson, L. (1977) 'Effect of the violation of assumptions upon significance levels of Pearson's r', *Psychological Bulletin,* 84: 373–7.

Hubbard, P., Kitchen, R., Bartley, B. and Fuller, D. (2002) *Thinking Geographically,* Continuum: London.

Kennedy, B.A. (1992) 'First catch your hare . . . research designs for individual projects', in A. Rogers, H. Viles and A.S. Goudie (eds) *The Students' Companion to Geography,* Blackwell: Oxford, pp. 128–34.

Kirkby, M.J., Naden, P.S., Burt, T.P. and Butcher, D.P. (1993) *Computer Simulation in Physical Geography,* 2nd edn, John Wiley & Sons: Chichester.

Kitchen, R. and Tate, N.J. (2000) *Conducting Research into Human Geography,* Prentice Hall: Harlow.

Knight, P.G. (1989) 'Stacking of basal debris layers without bulk freezing-on: isotopic evidence from West Greenland', *Journal of Glaciology,* 35 (120): 214–16.

Knight, P.G. and Parsons, T. (2003) *How to do your Essays, Exams and Coursework in Geography and Related Disciplines,* Nelson Thornes: Cheltenham.

Koch, G.S. Jr. and Link, R.F. (1970) *Statistical Analysis of Geological Data,* Dover Publications: New York.

Limb, M. and Dwyer, C. (eds) (2001) *Qualitative Methodologies for Geographers,* Arnold: London.

Mercer, C. (1999) 'Reconceptualizing state-society relations in Tanzania: are NGOs "making a difference"?', *Area,* 31: 247–58.

152

REFERENCES

M.H.R.A. (1996) *Modern Humanities Research Association Style Book*, 5th edn, M.H.R.A.: London.

Parry, J.T. (1960) 'The limestone pavements of northwestern England', *Canadian Geographer*, 16: 14–21.

Pigott, C.D. (1965) 'The structure of limestone surfaces in Derbyshire', *Geographical Journal*, 131: 41–4.

Phillips, I.L. Jr. (1971) *How to Think About Statistics*, Freeman: New York.

Poesen, J.W.A. and Torri, D. (1989) 'Mechanisms governing incipient motion of ellipsoidal rock fragments in concentrated overland flow', *Earth Surface Processes and Landforms*, 14: 469–80.

Popper, K.R. (1959) *The Logic of Scientific Discovery*, Hutchinson: London.

Price, R. (1989) *Scotland's Golf Courses*, Aberdeen University Press: Aberdeen.

Rhoads, B. L. and Thorn, C.E. (eds) (1996) *The Scientific Nature of Geomorphology*, John Wiley & Sons: Chichester.

Rose, L. and Vincent, P. (1986) 'Some aspects of the morphology of grikes: a mixture model approach', in K. Paterson and M.M. Sweeting (eds) *New Directions in Karst, Proceedings of the Anglo-French Karst Symposium*, September 1983, Geobook: Norwich, pp. 473–96.

Ryan, A. (1970) *The Philosophy of the Social Sciences*, Macmillan: London.

Shaw, G. and Wheeler, D. (1994) *Statistical Techniques in Geographical Analysis*, 2nd edn, David Fulton Publishers: London.

Sides, C.H. (1992) *How to Write and Present Technical Information*, Cambridge University Press: Cambridge.

Siegal, S. (1956) *Nonparametric Statistics for the Behavioral Sciences*, McGraw-Hill: New York.

Thomas, R.W. and Huggett, R.J. (1980) *Modelling in Geography*, Harper & Row: London.

Wainwright, J. and Mulligan, M.(eds) (2004) *Environmental Modelling*, John Wiley & Sons: Chichester.

Wainwright, J., Parsons, A.J. and Abrahams, A.D. (1995) 'A simulation study of the role of raindrop erosion in the formation of desert pavements', *Earth Surface Processes and Landforms*, 20: 277–91.

Willett, W.C., Stampfer, M.J., Manson, J.B., *et al.* (1993) 'Intake of *trans* fatty acids and risk of coronary heart disease among women', *The Lancet*, 341: 581–5.

Williams, P.W. (1966) 'Limestone pavements with special reference to western Ireland', *Institute of British Geographers, Transactions*, 40: 155–71.

INDEX